The Battle of Point Pleasant

A Battle of the Revolution
October 10th, 1774

Biographical Sketches of the Men Who Participated

By

Mrs. Livia Nye Simpson-Poffenbarger

Originally Published by the State Gazeteer,
Point Pleasant, West Virginia, 1909.

Reformatted and published by Jessica Duke,
Las Vegas, Nevada, 2011.

Book Layout and Cover Design by Jessica Duke.

Printed in the United States of America.
ISBN: 978-1-61393-014-4

The illustrations from the original text were retained and included in this printing. There were numerous spelling errors in the original text which have been corrected; however, spelling and grammatical errors in quotations and the like have been preserved in their original form.

DEDICATION

This little volume is dedicated to the memory of the brave colonists who, successful at the battle of Point Pleasant, had fought the opening-battle of the Revolution, in preserving the right arm of Virginia for the struggle with the Mother Country; thus making possible the blessings of liberty we now enjoy as a Nation.

MRS. LIVIA NYE SIMPSON-POFFENBARGER.

BATTLE OF POINT PLEASANT, OCTOBER 10th, 1774

TABLE OF CONTENTS.

BATTLE OF POINT PLEASANT.

Andrew Lewis, who commanded the colonial troops in the Battle of Point Pleasant, October 10, 1774, was the son of John Lewis and Margaret Lynn Lewis, his wife.

John Lewis was of Scotch Irish descent, having been born in France, 1673, where his ancestors had taken refuge from the persecution following the assassination of Henry IV. He married Margaret Lynn, the daughter of the "Laird of Loch Lynn, "of Scotland, and emigrated to Ireland, thence to America in 1729, and became the founder of Staunton, Virginia. Here, he planted a colony and reared a family that have given luster to American History.

Governor Gooch, of Williamsburg, then the seat of Government of Virginia, was the personal friend of Mrs. Lewis' father and hence granted her sons, together with one Benjamin Burden a land warrant for 500,000 acres of land in the James and Shenandoah Valleys, with the proviso that they were to locate one hundred families within ten years. They induced their friends from Scotland and the north of Ireland, and the Scotch Irish of Pennsylvania, to emigrate to Augusta County, Virginia. In her diary, Mrs. Lewis says: "It sounded like the gathering of the clans to hear the names of these settlers viz: McKees, McCues, McCampbells, McClungs, McKouns, Caruthers, Stuarts, Wallaces, Lyles, Paxtons, Prestons and Grisbys."

We quote the following from the *Ohio Archaeological and Historical Quarterly*, July, 1903, pp. 288, 289, 290:

"When John Randolph said that Pennsylvania had produced but two great men Benjamin Franklin, of Massachusetts, and Albert Gallatin, of Switzerland-he possibly did not know that the best blood of his own State was that of the Scotch-Irish people who went down from Pennsylvania and settled in the Valley. He likely did not know that the great and good Dr. Archibald Alexander, the founder of Liberty Hall, now Washington and Lee University (so much loved by Washington,) the very seat of culture and power of the Shenandoah and James, the greatest factor of the State's prowess, was a Pennsylvanian. He possibly did not know that Dr. Graham, the first president of this institution, was from Old Paxtang; that many of the families whose names are in the pantheon of old Dominion achievement, the families that give Virginia her prominence, in the sisterhood of States, had their American origin in Pennsylvania in the Scotch-Irish reservoir of the Cumberland Valley the McDowells, the Pattersons, the McCormacks, Ewings, McCorcles, Prestons, McCungs, Craigs, McCulloughs, Simpsons, Stewarts, Moffats, Irwins, Hunters, Blairs, Elders, Grahams, Finleys, Trimbles, Rankins, and hundreds of others, whose achievements mark the pathway of the world's progress. John Randolph possibly did not know that when the first Declaration of Independence by the American patriots was issued by the member s of Hanover Church out there in Dauphin county, when on June 4th, 1774, they declared "that in the event Great Britain attempting to force unjust laws upon us by the strength of Arms, our cause we leave to heaven and our rifles." This declaration was certainly carried to Mecklenburg to give the sturdy people of that region inspiration for the strong document issued by them a year later, and which gave Jefferson a basis for the Declaration of 1776. There was much moving from Pennsylvania into Virginia and North Carolina before the Revolution, and Hanover Presbytery in the Valley was largely made up of people from Pennsylvania, whose petition of ten thousand names for a free church in a free land, made in 1785, was the force back of Jefferson's bill for religious tolerance, a triumph for freedom that has always been considered a Presbyterian victory by the Scotch-Irish of America.

We know that Dr. Sankey of Hanover Church was a minister in Hanover Presbytery, and that he was followed into Virginia by large numbers of- the Hanover congregation, who kept up a constant stream into the Valley. By the way, two settlements were arrangement with made by this congregation in Ohio. Col. Rogers, Gov. Bushnel's secretary, derives his descent from them. The population of North Carolina at the outbreak of the Revolution was largely made up of Scotch-Irish immigrants from Pennsylvania and the Virginia Valley who had a public school system before the war. These were the people who stood with the Rev. David Caldwell on the banks of the Alamance May 16th, 1771, and received the first volley of shot fired in the contest for Independence. This same blood coursed the veins of the patriot army with Lewis at Point Pleasant, the first battle of the Revolutionary War, fought October 10, 1774, Lord Dunmore having no doubt planned the attack by the Indians to discourage the Americans from further agitation of the then pending demand for fair treatment of the American Colonies at the hands of Great Britain. It was this blood that coursed the veins of those courageous people who, having survived the Kerr's creek massacre, were carried to a

Shawnee village in Ohio, and on being bantered to sing by the Indians in their cruel sport, sang Rouse's version of one of the Psalms. "Unappalled by the bloody scene," says the Augusta historian, "through which they had already passed, and the fearful tortures awaiting them, within the dark wilderness of forest, when all hope of rescue seemed forbidden; undaunted by the fiendish revelings of their savage captors, they sang aloud with the most pious ferver –

> "On Babel's stream we sat and wept wen
> Zion we thought on,
> In midst thereof we hanged our harps the
> willow trees among.
> For then a song required they who did us
> captive bring,
> Our spoilers called for mirth and said, a
> song of Zion sing."

It was this blood that fought the battle of King's Mountain, which victory gave the patriots the courage that is always in hope; it was the winning force at Cowpens, at Guilford, where Rev. Samuel Houston discharged his rifle fourteen times, once for each ten minutes of the battle. These brave hearts were in every battle of the Revolution, from Point Pleasant in 1774 to the victory of Wayne at the Maumee Rapids twenty years later, for the War of Independence continued in the Ohio Country after the treaty of peace. And yet, after all this awful struggle to gain and hold for America the very heart of the Republic, one of the gentlemen referred to by Mr. Randolph wrote pamphlets in which he derided as murderers the courageous settlers of our blood on the occasions they felt it necessary to "remove" Indians with their long rifles. After all the struggle, he too would have made an arrangement with England by which the Ohio river would have been the boundary line."

These were the people who in coming" to America had not only secured for themselves that personal religious freedom of a church without a Bishop and ultimately a state without a King, but they became recruits in the Army of Andrew Lewis, the hero of the Battle of Point Pleasant, and like many of their country-men, continued in the army, (those who had not met the fate of battle,) and became the flower of Virginia's Colonial Army.

THE STATUS OF THE BATTLE AT POINT PLEASANT

While the Battle of Point Pleasant has always been conceded to have been the most terrific conflict ever waged between the white man and the Indian, its full significance has not been made the text of American history. We quote however, from a few of the American writers, showing their estimate of it.

Roosevelt, in *The Winning of the West*, Vol. II, chap. 2, says: "Lord Dunmore's War, waged by Americans for the good of America, was the opening act in the drama whereof the closing scene was played at Yorktown. It made possible the two fold character of the Revolutionary War, wherein on the one hand the Americans won by conquest and colonization, new lands for their children, and on the other wrought out their national independence of the British King."

Kercheval's *History of the Valley*, p. 120, says: "Be it remembered, then, that this Indian war was but a portico to our revolutionary war, the fuel for which was then preparing, and which burst into a flame, the ensuing year. Neither let us forget that the Earle of Dunmore was at this time governor of Virginia; and that he was acquainted with the views and designs of the British Cabinet, can scarcely be doubted. What then, suppose ye, would be the conduct of a man possessing his means, filling a high, official station, attached to the British government, and master of consummate diplomatic skill."

Dr. John P. Hale, in writing of the Battle of Point Pleasant, says, in the *History of the Great Kanawha Valley*, Vol. I, pp. 114, 115, "Early in the spring of 1774, it was evident that the Indians were combining for aggressive action. * * * It was decided that an army of two divisions should be organized as speedily as practicable one to be commanded by Gen. Lewis, and the other by Lord Dunmore, in person. * * * Gen. Lewis 'army rendezvoused at Camp Union (Lewisburg) about September 1st, and was to March from there to the mouth of Kanawha; while against the colonies. Gov. Dunmore was to go the northwest route, over the Braddock trail, by way of Fort Pitt, and thence down the Ohio river and form a junction with Gen. Lewis at the mouth of Kanawha.

The aggregate strength of this southern division of the army was about eleven hundred; the strength of the northern division, under Lord Dunmore, was about fifteen hundred. On the 11th of September Gen. Lewis broke camp, and, with Captain Matthew Arbuckle, an intelligent and experienced frontiersman, as pilot, marched through a pathless wilderness. They reached Point Pleasant on the 30th day of September, after a fatiguing march of nineteen days. Gen. Lewis for several days anxiously awaited the arrival of Lord Dunmore, who, by appointment, was to have joined him here on the 2nd of October. Having no intelligence from him, Lewis dispatched messengers up the Ohio river to meet him, or learn what had become of him.

Before his messengers returned, however three messengers (probably McCulloch, Kenton and Girty) arrived at his camp on Sunday, the 9th of October, with orders from Lord Dunmore to cross the river and meet him before the Indian towns in Ohio. This is, substantially, the current version of matters: but authorities differ.

Some say the messenger arrived on the night of the 10th, after the battle was fought; others say they did not arrive until the 11th, the day after the battle, and Col. Andrew Lewis, son of Gen. Andrew Lewis, says his father received no communication whatever from Lord Dunmore after he (Lewis) left camp Union, until after the battle bad been fought, and Lewis of his own motion, had gone on into Ohio, expecting to join Dunmore and punish the Indians, when he received an order to stop and return to the Point. This order (by messenger) Lewis disregarded, when Lord Dunmore came in person, and after a conference and assurances from Dunmore that he was about negotiating a peace, Lewis reluctantly retraced his steps. In the very excited state of feeling, then existing between the colonies and the mother country. It was but natural that the sympathies of Lord Dunmore, a titled English nobleman, and holding his commission as governor of Virginia at the pleasure of the crown, should be with his own country; but it was not only strongly suspected, but generally charged, that, while he was yet acting as governor of Virginia, and before he had declared himself against the colonies, he was unfairly using his position and influence to the prejudice of his subjects.

According to the account of Col. Stewart, when the interview was over between Gen. Lewis and the messengers of Lord Dunmore, on the 9th, Lewis gave orders to break camp at an early hour next morning, cross the river, and take up their march towards the Indian towns; but the fates had decreed otherwise. At the hour for starting, they found themselves confronted by an army of Indian braves, eight hundred to one thousand strong, in their war paint, and commanded by their able and trusted leaders, Cornstalk, Logan, Red Hawk, Blue Jacket and Elinipsico, and some authors mention two or three others. Instead of a hard day's marching, Lewis army had a harder day's fighting the important, desperately contested, finally victorious, and ever-memorable battle of Point Pleasant. No "official report" of this battle has been preserved, or was ever written, so far as can be learned. There are several good reasons, apparently, for this omission. In the first place, the time, place and circumstances were not favorable for preparing a formal official report. In the second place, Lord Dunmore, the superior officer, to whom Gen. Lewis should, ordinarily, have reported, was himself in the field, but a few miles distant, and Gen. Lewis was expecting that the two divisions of the army would be united within a few days; and, in the third place, the "strained relations" between the colonies and the mother country were such, and the recent action of Gov. Dunmore so ambiguous, that Gen. Lewis was probably not inclined to report to him at all."

The same author, in the same volume, at pages 122, 128, 129, 130, 131 and 132, says: "Col. Stewart, one of the first to write about the battle, after Arbuckle's short account, was himself present, was well known to Gen. Lewis (and a relative by marriage), says Gen. Lewis received a message from Gov. Dunmore, on the 9th, telling him to cross the Ohio and join him. Burk and others, say the messengers came after the battle, and mention Simon Kenton and Simon Girty among the messengers. Col. Andrew Lewis says his father received no communication of any sort from Gov. Dunmore, until ordered to return from Ohio. * * * * It has been stated that there were not only suspicions, but grave charges, that Governor Dunmore acted a double part, and that he was untrue and treacherous to the interests of the colony he governed. As he is inseparably connected with the campaign (often called the Dunmore War), and its accompanying history, and the inauguration of the Revolution, it may be well to briefly allude to his official course just before, during and after the campaign that his true relations to it, and to the colony, may be understood; and, also, to show that the "Revolution" was really in progress; that this campaign was one of the important early moves on the historical chessboard, and that the battle of Point Pleasant was, as generally claimed, the initiatory battle of the great drama. In the summer of 1773, Governor Dunmore made, ostensibly, a pleasure trip to Fort Pitt; here he established close relations with Dr. Connally, making him Indian agent, land agent, etc. Connally was an able active and efficient man, who thereafter adhered to Dunmore and the English cause. It is charged that Connally at once began fomenting trouble and ill-feeling between the colonies of Virginia and Pennsylvania in regard to the western frontier of Pennsylvania, then claimed by both colonies, but held by Virginia, hoping by such course to prevent the friendly co-operation of these colonies against English designs; and, also to incite the Indian tribes to resistance of western white encroachments upon their hunting grounds, and prepare the way forgetting their co-operation with England against the colonies, when the rupture should come. In December, 1773, the famous "cold- water tea" was made in Boston harbor. In retaliation the English government blockaded the port of Boston, and moved the capital of the colony to Salem. When this news came, in 1774, the Virginia assembly, being in session, passed resolutions of sympathy with Massachusetts, and strong disapproval of the course of England; whereupon Governor Dunmore peremptorily dissolved the assembly. They met privately, opened correspondence with the other colonies, and proposed co-operation and a colonial congress. On the 4th of September, 1774, met, in Philadelphia, the first continental congress Peyton Randolph, of Virginia, president; George Washington, R. H. Lee, Richard Bland, Patrick Henry, Benjamin Harrison and Edmund Pendleton members from Virginia. They passed strong resolutions; among others; to resist taxation and other obnoxious measures; to raise minute men to forcibly resist coercion; and, finally resolved to cease all official intercourse with the English government. In the meantime, Dr. Connally had been carrying out the programme of the northwest. He had taken possession of the fort at Fort Pitt, and renamed it Fort Dunmore; was claiming lands under patents from Governor Dunmore, and making settlements on them; had been himself arrested and imprisoned for a time by Pennsylvania; had the Indian tribes highly excited, united in a strong confederacy and threatened war; then came the massacre of Indians above Wheeling, at Capitina and at Yellow creek, said to have grown out of Connally's orders. While the continental congress was passing the resolutions above mentioned, and which created a breach between the colonies and the mother country past healing, Governor Dunmore and General Lewis were organizing and marching their armies to the west. Instead of uniting the forces into one army, and marching straight to the Indian towns and conquering or dictating a lasting peace, Lord Dunmore took the larger portion of the army by a long detour by Fort Pitt, and thence down the Ohio, picking up on the way Dr. Connally and Simon Girty, whom he made useful. At Fort Pitt, it is said, he had held a conference with some of the Indian chiefs, and came to some understanding with them, the particulars of which are not known. Instead of uniting with Lewis at the mouth of Kanawha, as had been arranged, but which was probably not intended, he struck off from the Ohio river at the mouth of Hockhocking and marched for the Indian towns on the Pickaway plains, without the support of Lewis army, delaying long enough for the Indians to have annihilated Lewis division if events had turned out as Cornstalk had planned. He (Cornstalk) said it was first their intention to attack the "Long Knives" and destroy them, as they crossed the river, and this plan would have been carried out, or attempted, but for the long delay of Lewis' awaiting the arrival of Lord Dunmore. They afterwards, upon consultations, changed their plans, and determined to let Lewis cross the river and then ambush him somewhere near their own homes, and farther from his (Lewis') base; but the Indians had no organized commissary or transportation arrangements, and could only transport such amount of food as each brave could carry for his own sustenance; this was necessarily, a limited amount, and Lewis' delay in crossing had run their rations so short that they were obliged to cross, themselves, and force a fight, or break camp and go to hunting food. They crossed in the night, about three miles above the Point, on rafts previously constructed, and expected to take Lewis' army by surprise; and it will be seen how near they came to accomplishing it. It was prevented by the accident of the early hunters, who were out before daylight, in violation of orders.

Dr. Campbell says there was considerable dissatisfaction in Lewis camp, for some days before the battle growing out of the manner of serving the rations, and especially the beef rations; the men claimed that the good and bad beef were not dealt out impartially. On the 9th, Gen. Lewis ordered that the poorest beeves be killed first, and distributed to all alike. The beef was so poor that the men were unwilling to eat it, and, although it was positively against orders to leave camp without permission, about one hundred men started out before day, next morning (the 10th), in different directions, to hunt and provide their own meat. Many of these did not get back, nor know of the battle until night, when it was all over. This was a serious reduction of the army at such a time.

Col. Andrew Lewis (son of General Andrew,) in his account of the Point Pleasant campaign, says: “It is known that Blue Jacket, a Shawnee Chief, visited Lord Dunmore's camp, on the 9th, the day before the battle, and went straight from there to the Point, and some of them went to confer with Lord Dunmore immediately alter the battle.” It is also said that Lord Dunmore, in conversation with Dr. Connally, and others, on the 10th, the day of the battle, remarked that “Lewis is probably having hot work about this time.”

When Lewis had crossed the river, after the battle, and was marching to join Dunmore, a messenger was dispatched to him twice in one day, ordering him to stop and retrace his steps the messenger in each instance, being the afterward notorious Simon Girty. Gen. Lewis bad, very naturally, become much incensed at the conduct of Lord Dunmore, and took the high-handed responsibility advised and sanctioned by his officers and men of disobeying the order of his superior in command, and boldly marching on towards his camp. When within about two and one-half miles of Lord Dunmore's headquarters, which he called Camp Charlotte, after Queen Charlotte, wife of his majesty, George III., he came out to meet Lewis in person, bringing with him Cornstalk, White Eyes (another noted Shawanee chief), and others, and insisted on Lewis's returning as he (Dunmore) was negotiating a treaty of peace with the Indians. He sought an introduction to Lewis' officers, and paid them some flattering compliments, etc. Evidently it did not comport with Lord Dunmore's plans to have Gen. Lewis present at the treaty, to help the negotiation by suggestions, or to have the moral support of his army to sustain them. So much did Lewis' army feel the disappointment and this indignity, that Col. Andrew, his son, says that it was with difficulty Gen. Lewis could restrain his men (not under very rigid discipline, at best) from killing Lord Dunmore and his Indian escort. But the result of the personal conference was that Gen, Lewis, at last with the utmost reluctance of himself and army, consented to return, and to disband his army upon his arrival at Camp Union, as ordered.

Suppose Lewis had attempted to cross the river, and been destroyed, or had crossed and been ambushed and demolished in the forest thickets of Ohio, or that Cornstalk had succeeded, as he came so near doing, in surprising him in his own camp, on the morning of the 10th, or after that; suppose the Indians bad succeeded in turning the so evenly balanced scale in their favor, during the fight, as they came so near doing, and had annihilated Lewis' army, as they might have done, having them penned up in the angle of two rivers, who can doubt in view of all the facts above noted, that Lord Dunmore would have been responsible for the disaster? Who can doubt, as it was, that he was responsible for the un- necessary sacrifice of life, at the Point, on the 10th? Who can doubt that, with the two divisions of the army united, as per agreement, and Lord Dunmore and Lewis acting in unison and good faith, they could have marched to the Indian towns, and utterly destroyed them, or dictated a favorable and lasting peace, and maintained it as long as they pleased, by holding important hostages? But, clearly, the policy of the governor was dictated by ulterior and sinister motives; his actions were not single-minded. Col.

Andrew Lewis says: "It was evidently the intention of the old Scotch villain to cut off Gen. Lewis' army.' Burk the historian, says: "The division under Lewis was devoted to destruction, for the purpose of breaking the spirit of the Virginians." Withers, Doddridge, and others, express the same views. Gen. Lewis and his army were convinced of the fact; Col. Stewart bad no doubt of it, and nearly every one who has written on the subject has taken the same view of it. A few only are willing to give him the benefit of a doubt. If this design to destroy Lewis' army had succeeded, it is almost certain that the English, through Lord Dunmore, would have perfected an alliance, offensive and defensive, with the victorious Indians, against the colonies, and every white settlement west of the Alleghenies would probably have been cut off. It would have been difficult or impossible, for a time, to raise another army for the defense of the western border; the tory element would have been encouraged and strengthened, the revolutionary element correspondingly discouraged, the rebellion (?) crushed, and Lord Dunmore would have been the hero of the age. Upon what slender and uncertain tenures bang the destinies of nations, and the fate of individuals! The closely-won success of Lewis was not only an immediate victory over the Indians, but a defeat of the machinations of the double dealing governor, and the projected Anglo-Indian alliance. If this view of it is established the claim of the battle of Point Pleasant as being the initiatory battle of the revolution; and, although small in itself, when its after results and influences are considered it stands out in bold relief as one of the important and decisive victories of history. A few words more and we shall be done with Lord Dunmore. Upon his return to Williamsburg, the Assembly, upon his own ex-parte statement of the results of the campaign, passed a vote of thanks for his "valuable services,' etc., which, it is said, they very much regretted when they learned more of the facts. Just after the battle of Lexington (April 19, 1775), he had all the powder that was stored in the colonial magazine at Williamsburg secretly conveyed on board an armed English vessel lying off Yorktown, and threatened to lay Williamsburg in ashes at the first sign of insurrection. Patrick Henry raised a volunteer force to go down and compel him (Dunmore) to restore the powder; but as this was impracticable, he agreed to pay, and did pay for it, and then issued a proclamation declaring "One Patrick Henry and his followers rebels.' He had previously threatened Thomas Jefferson with prosecution for treason, and had commenced proceedings. About this time, having previously sent his family on an English naval vessel, he made his own escape, by night, to the English fleet and commenced a system of depredations along the coast, burning houses, destroying crops, etc. He tried to bring his scheme of Indian co-operation to bear, and sent a message to his old friend, Connally, with a commission as Colonel, and instructed him to secure the co-operation of as many of the western militia commanders as possible, by large rewards; to form an alliance with the Indians, collect his forces at Fort Pitt, and march through Virginia and meet him. Fortunately, Col. Connally was captured and imprisoned, and the scheme exposed and thwarted He (Dunmore) issued a proclamation granting freedom, to all the slaves who would flock to his standard, and protection to the Tories. Among other acts of violence, he burned Norfolk, the then largest and most important town in Virginia. Upon his flight, the Assembly met and declared his office vacant, and proceeded to fill it; and, for the first time, Virginia had entire "home rule". Upon the petition of citizens of Dunmore county, which had been named in his honor, the name was abolished, and the county called Shenandoah. In 1776, Lord Dunmore and his fleet and hangers-on were at Guynne's Island, in the Chesapeake Bay, where, as an interesting example of poetic or retributive justice, Gen. Lewis in command of the Virginia troops, attacked, defeated, and drove them off, with heavy loss, Gen. Lewis himself, firing the first gun, soon after which the ex-Governor, a sadder and wiser man "left the country for the country's good."

It will thus be seen that Dunmore, the Tory Governor of Virginia, knew that the war of the Revolution was inevitable. John Adams dates the opening of the Revolution in 1760. The people had tired of taxation without representation. In 1764 we find an organized opposition to oppressive taxation in Boston. In 1765, was passed the Stamp Act and in that year was organized the Sons of Liberty. In 1766 the Royal Artillery was in Boston. In 1767, a duty was imposed on tea. In 1768 British troops were sent to Boston. In 1768 in Virginia was passed the non-importation agreement, followed in 1770 by the Boston Massacre.

In the Parliament of England, the discussion of the taxation of the colonies did not tend to allay their determination to thwart all oppression and when George III determined at all odds to impose taxation the matter was settled in the heart of every loyal American, whether the vow was expressed or implied. It is well authenticated that, to occupy the attention of the colonial forces that they might not have so much (time in which to brood over the oppression of the mother country,) it was necessary to incite the Indians to attack the frontiers and so divert the attention of the colonists from their quarrel with the mother country and at the same time impress upon them a feeling of dependence upon British arms and means for the safety of their lives and homes. One of the quickest to avail himself of this method of resisting the onflowing tide of this demand for Liberty was Governor Dunmore.

Virginia had been the first in 1764 to pass a Resolution, defying the British authority as is seen by the following, introduced by Patrick Henry, in the House of Burgesses, and carried:

"Resolved, therefore, That the General Assembly of this colony, together with his majesty or substitute, have, in their representative capacity, the only exclusive right and power to lay taxes and impositions upon the inhabitants of this colony; and that every attempt to vest such power in any person or persons whatsoever, other than the General Assembly aforesaid, is illegal, unconstitutional, and unjust, and has a manifest tendency to destroy British, as well as American, Freedom."

In this same year 1764, Patrick Henry originated the great question which led to the final independence of the United States.

When, in January, 1765, the famous stamp act was passed that for a while stunned the whole country, and confounded the people, it was Virginia, led by the matchless Henry, that stood forth to raise the drooping spirits of the colonists, and it is said his election to the house of burgesses was with express reference to his opposition to the stamp act, and the adoption of a series of resolutions in 1765, chief among which was the one above referred to.

Upon the death of Mr. Henry, in his private papers, was found the original manuscript, embracing the above Resolution with others, bearing the following narrative, written on the back of it by Mr. Henry, himself:

"The within resolutions passed the house of burgesses in May, 1765. They formed the first opposition to the stamp act, and the scheme of taxing America by the British parliament.' All the colonies, either through fear, or want of opportunity to form an opposition, or from influence of some kind or other, had remained silent. I had been for the first time elected a burgess, a few days before, was young, inexperienced, unacquainted with the forms of the house, and the members that composed it. Finding the men of weight averse to opposition, and the commencement of the tax at hand, and that no person was likely to step forth, I determined to venture, and alone, unadvised, and unassisted, on a blank leaf of an old law book wrote the within. Upon offering them to the house, violent debates ensued. Many threats were uttered, and much abuse cast upon me, by the party for submission. After a long and warm contest, the resolutions passed by a very small majority, perhaps of one or two only. The alarm spread through America with astonishing quickness, and the ministerial party were overwhelmed. The great point of resistance to British taxation was universally established in the colonies. This brought on the war, which finally separated the two countries, and gave independence to ours."

The Virginia house of burgesses continued to pass resolutions of defiance until the session of 1768-9, when the house was dissolved by the governor. This house had the merit of originating that powerful engine of resistance, corresponding committees between the legislatures of the colonies, a measure so nearly coeval in the two states of Virginia and Massachusetts that it would have been, at that time with their slow methods of communication, impossible to have borrowed the idea one from the other; so that they are equally entitled to that honor, although Mrs. Warren, a Massachusetts historian of that time, admits that the measure originated in Virginia.

It will thus be seen that when the colonists met in Congress in Philadelphia, September 4, 1774, that all over Virginia it was believed, as Patrick Henry had so eloquently asserted, that the war was inevitable, and the people were ready to voice his sentiment, "Let it come." Considering all these facts, we can well credit Howe, the Virginia historian who says, "While Virginia was employed in animating her sister states to resistance, her governor was employed in the ignoble occupation of fomenting jealousies and feuds between the province, which it should have been his duty to protect from such a calamity, and Pennsylvania, by raising difficult questions of boundary, and exciting the inhabitants of the disputed territory to forswear allegiance to the latter province; hoping thus, by affording a more immediately exciting question, to draw off the attention of these too important provinces from the encroachments of Great Britain. This scheme, as contemptible as it was iniquitous, wholly failed, through the good sense and magnanimity of the Virginia council. Lord North, full of his feeble and futile schemes of cheating the colonies out of their rights, took off the obnoxious duties with the exception of three pence per pound on tea; and, with the ridiculous idea that he might fix the principle upon the colonies by a precedent, which should strip it of all that was odious, offered a draw-back equal to the import duty. This induced the importation of tea into Boston harbor which, being thrown overboard by some of the citizens, called down upon their city all the rigor of the celebrated Boston port bill.

A draft of this bill reached the Virginia legislature while in session; an animated protest, and a dissolution of the assembly by the governor, of course followed. On the following day the members convened in the Raleigh tavern and, in an able and manly paper, expressed to their constituents and their government those sentiments and opinions which they had not been allowed to express in a legislative form.

This meeting recommended a cessation of trade with the East India Company, a Congress of deputies from all the colonies, 'declaring their opinion, that an attack upon one of the colonies was an attack upon all British in America, ' and calling a convention of the people of Virginia. The sentiments of the people accorded with those of their late delegates; they elected members who met in convention at Williamsburg, on the 1st of August, 1774.

This convention went into a detailed view of their rights and grievances, discussed measures of redress for the latter, and declared their determination never to relinquish the former; they appointed deputies to attend a general Congress, and they instructed them how to proceed. The Congress met in Philadelphia, on the 4th of September, 1774.

While Virginia was engaged in her efforts for the general good, she was not without her peculiar troubles at home. The Indians had been for some time waging a horrid war upon the frontiers, when the indignation of the people at length compelled the reluctant governor to take up arms, and march to suppress the very savages he was thought to have encouraged and excited to hostility by his intrigues.

Lord Dunmore marched the army in two divisions: the one under General Andrew Lewis he sent to the junction of the Great Kanawha with the Ohio, while he himself marched to a higher point on the latter river, with pretended purpose of destroying the Indian towns and joining Lewis at Point Pleasant; but it was believed with the real object of sending the whole Indian force to annihilate Lewis' detachment, and thereby weaken the power and break down the spirit of Virginia. If such was his object he was signally defeated through the gallantry of the detachment, which met and defeated the superior numbers of the enemy at Point Pleasant, after an exceeding hard-fought day, and the loss of nearly all its officers. The day after the victory, an express arrived from Dunmore with orders for the detachment to join him at a distance of 80 miles, through an enemy's country, without any conceivable object but the destruction of the corps. As these orders were given without knowledge of the victory, Col. Lewis was proceeding to the destruction of the Shawanee villages, when he was informed the governor had made peace.

Another evidence of Dunmore's intention to have the army of Andrew Lewis destroyed at Point Pleasant, is found in Kercheval's *History of the Valley*, p. 118, as follows: "It was the general belief among the officers of our army, at the time, that the Earl of Dunmore, while at Wheeling, received advice from his government of the probability of the approaching war between England and the colonies, and that afterwards, all his measures, with regard to the Indians, had for their ultimate object an alliance with those ferocious warriors for the aid of the mother country in their contest with us. This supposition accounts for his not forming a junction with the army of Lewis at Point Pleasant. This deviation from the original plan of the campaign jeopardized the army of Lewis and well nigh occasioned its total destruction. The conduct of the Earl at the treaty, shows a good understanding between him and the Indian chiefs. He did not suffer the army of Lewis to form a junction with his own, but sent them back before the treaty was concluded, thus risking the safety of his own forces, for at the time of the treaty, the Indian warriors were about his camp in force sufficient to have intercepted his retreat and destroyed his whole army."

Again, Kercheval says: "We now proceed to examine the question, how far facts and circumstances justify us in supposing the Earl of Dunmore himself was instrumental in the Indian war of 1774. It has already been remarked that this Indian war was but the precursor to our revolutionary war of 1775 that Dunmore the then governor of Virginia, was one of the most inveterate and determined enemies to the revolution that he was a man of high talents, especially for intrigue and diplomatic skill that occupying the station of commander-in-chief of the large and respectable State of Virginia, he possessed means and power to do much to serve the views of Great Britain. And we have seen, from the preceding pages, how effectually he played his part among the inhabitants of the western country. I was present myself when a Pennsylvania magistrate, of the name of Scott, was taken into custody, and brought before Dunmore, at Prestone old Fort; he was severely threatened and dismissed, perhaps on bail, but I do not recollect how; another Pennsylvania magistrate was sent to Staunton jail. And I have already shown in the preceding pages, that there was a sufficient preparation of materials for this war in the predisposition and hostile attitude of our affairs with the Indians; that it was consequently no difficult matter with a Virginia governor to direct the incipient state of things to any point most conclusive to the grand end he had in view, namely, in weakening our national strength in some of the best and most efficient parts. If, then, a war with the Indians might have a tendency to produce this result, it appears perfectly natural and reasonable to suppose that Dunmore would make use of the power and influence to promote it, and although the war of 1774 was brought to a conclusion before the year was out, yet we know that this fire was scarcely extinguished before it burst into a flame with tenfold fury, and two or three armies of the whites were sacrificed before we could get the Indians subdued; and this unhappy state of our affairs with the Indians happening during the severe conflict of our revolutionary war, had the very effect, I suppose, Dunmore had in view namely, dividing our forces and enfeebling our aggregate strength; and that the seeds of these subsequent wars with the Indians were sown in 1774 and 1775, appears almost certain.

"And the first we shall mention is a circular sent by Maj. Connoly, his proxy, early in the spring of the year 1774, warning the inhabitants to be on their guard the Indians were very angry, and manifested so much hostility, that he was apprehensive they would strike some where as soon as the season would permit, and enjoining the inhabitants to prepare and retire into Forts, &c. It might be useful to collate and compare this letter with one he wrote to Capt. Cresap on the 14th of July following; see hereafter. In this letter he declares there is a war or danger of war, before the war is properly begun; in that to Capt. Cresap, he says, "the Indians deport themselves peaceably;" when Dunmore and Lewis and Cornstalk we are all out on their march for battle.

"This letter produced its natural result. The people fled into Forts, and put themselves into a posture of defense, and the tocsin of war resounded from Laurel Hill to the banks of the Ohio river. Capt. Cresap who was peaceably at this time employed in building houses and improving lands, on the Ohio River, received this letter, accompanied, it is believed, with a confirmatory message from Col. Croghan and Maj. McGee, Indian agents and interpreters; and he thereupon immediately broke up his camp, and ascended the River to Wheeling fort, the nearest place of safety from whence it is believed he intended speedily to return home; but during his stay at this place, a report was brought to the Fort that two Indians were coming down the River. Capt. Cresap, supposing from every circumstances, and the general aspect of affairs, that war was inevitable, and in fact already begun, went up the River with his party; and two of his men, of the name of Chenoweth and Brothers, killed, these two Indians. Beyond controversy this is the only circumstance in the history of this Indian war, in which his name can in the remotest degree be identified with any measure tending to produce this war; and it is certain that the guilt or innocence of this affair will appear from this date. It is notorious, then, that those Indians were killed not only after Capt. Cresap had received Connoly's letter, and after Butler's men were killed in the canoe, but also alter the affair at Yellow Creek, and after the people had fled into the Forts."

The same author further says, on pages 128-130, inclusive,

"The Governor of Virginia, whatever might have been his views as to the ulterior measures, lost no time in preparing to meet this storm. He sent orders immediately to Col. Andrew Lewis, of Augusta county, to raise an army of about one thousand men, and to march with all expedition to the mouth of the Great Kanawha, on the Ohio River, where, or at some other point, he would join him, after he had got together another army, which he intended to raise in the northwestern counties, and command in person. Lewis lost no time, collected the number of men required, and marched without delay to the appointed place of rendezvous.

"But the Earl was not quite so rapid in his movements, which circumstance the eagle eye of old Cornstalk, the general of the Indian army, saw, and was determined to avail himself of, foreseeing that it would be much easier to destroy two separate columns of an invading army before than after their junction and consolidation. With this view he marched with all expedition to attack Lewis before he was joined by the Earle's army from the north, calculating, confidently no doubt, that if he could destroy Lewis, he would be able to give a good account of the army of the Earl.

"The plan of Cornstalk appears to have been those of a consummate and skillful general, and the prompt and rapid execution of them displayed the energy of a warrior. He, therefore, without loss of time, attacked Lewis at his post. The attack was sudden, violent, and I believe unexpected. It was nevertheless well fought, very obstinate, and of long continuance; and as both parties fought with rifles, the conflict was dreadful; many were killed on both sides, and the contest was only finished with the approach of night. The Virginians, however, kept the field, but lost many able officers and men, and among the rest, Col. Charles Lewis, brother to the commander-in-chief.

This battle of Lewis' opened an easy and unmolested passage for Dunmore through the Indian country; but it is proper to remark here, however, that when Dunmore arrived with his wing of the army at the mouth of the Hockhocking River, he sent Capt. White-eyes, a Delaware chief, to invite the Indians to a treaty, and he remained stationary at that place until White-eyes returned, who reported that the Indians would not treat about peace. I presume, in order of time, this must have been just before Lewis' battle; because it will appear in the sequel of this story, that a great revolution took place in the minds of the Indians after the battle.

"Dunmore, immediately upon the report of White-eyes that the Indians were not disposed for peace, sent an express to Col. Lewis to move on and meet him near Chillicothe, on the Scioto river, and both wings of the army were put in motion. But as Dunmore approached the Indian town, he was met by flags from the Indians, demanding peace, to which he acceded, halted his army, and runners were sent to invite the Indian chiefs, who cheerfully obeyed the summons, and came to the treaty save only Logan, the great orator, who refused to come. It seems, however, that neither Dunmore nor the Indian chiefs considered his presence of much importance, for they went to work and finished the treaty without him referring, I believe, some unsettled points for future discussion, at a treaty to be held the ensuing summer or fall at Pittsburg. This treaty, the articles of which I never saw, nor do I know that, they were ever recorded, concluded Dunmore's war, in September or October, 1774. After the treaty was over, old Cornstalk, the Shawnee chief, accompanied Dunmore's army until they reached the mouth of the Hockhocking on the Ohio River; and what was more singular, rather made his home in Capt. Cresap's tent, with whom he continued on terms of the most friendly familiarity. I consider this circumstance as positive proof that the Indians themselves neither considered Capt. Cresap the murderer of Logan's family, nor the cause of the war. It appears, also, that at this place the Earl of Dunmore received dispatches from England. Dodridge says he received these on his march out.

But we ought to have mentioned in its proper place, that after the treaty between Dunmore and the Indians commenced near Chillocothe, Lewis arrived with his army, and encamped two or three miles from Dunmore, which greatly alarmed the Indians, as they thought he was so much irritated at losing so many men in the late battle that he would not easily be pacified; nor would they be satisfied until Dunmore and old Cornstalk went into Lewis' camp to converse with him.

Dr. Doddridge represents this affair in different shades of light from this statement. I can only say I had my information from an officer who was present at the time.

But it is time to remind the reader, that, although I have wandered into such a minute detail of the various occurrences, facts and circumstances of Dunmore's war; and all of which as a history may be interesting to the present and especially to the rising generation; yet it is proper to remark that I have two leading objects chiefly in view first, to convince the world, that whoever might be the cause of the Indian war in 1774, it was nut Capt. Cresap; secondly, that from the aspect of our political affairs, at that period, and from the known hostility of Dunmore to the American Revolution, and withal to the subsequent conduct of Dunmore, and the dreadful Indian war that commenced soon after the beginning of our war with Great Britain I say, from all these circumstances, we have infinitely stronger reasons to suspect Dunmore than Cresap; and I may say that the dispatches above mentioned that were received by Dunmore at Hockhocking, although after the treaty, were yet calculated to create suspicion.

But if, as we suppose, Dunmore was secretly at the bottom of this Indian war, it is evident that he could not with propriety appear personally in a business of this kind; and we have seen and shall see, how effectually his sub-governor played his part between the Virginians and Pennsylvanians; and it now remains for us to examine how far the conduct of this man (Connolly) will bear us out in the supposition that there was also some foul play, some dark intriguing work to embroil the western country in an Indian war."

Hon. V. A. Lewis who is the author of the History of the Virginias compiled in Hardesty's *Historical and Geographical Encylopedia* of 1883 pays the following tribute to the Battle of Point Pleasant:

"To the student of history no truth is more patent than this, that the battle of Point Pleasant, was the first in the series of the Revolution, the flames of which were being kindled by the oppression of the mother country and the resistance of the same by the feeble but determined colonies. It is a well known fact that the emissaries of Great Britain were then inciting the Indians to hostilities against the frontier for the purpose of distracting attention, and thus preventing the consummation of the Union which was then being formed to resist the tyranny of their armed oppression. It is also well known that Lord Dunmore was an enemy of the colonists, by his rigid adherence to the royal cause and his efforts to induce the Indians to co-operate with the English, and thus assist in reducing Virginia to subjection. It has been asserted that he intentionally delayed the progress of the left wing of the army that the right might be destroyed at Point Pleasant. Then at the mouth of the Great Kanawha river on the 19th (10th) day of October, 1774, there went whizzing through the forest the first volley of a struggle for liberty, which, in the grandeur and importance of its results, stands without parallel in the history of the world. On that day the soil upon which Point Pleasant, now stands drank the first blood shed in defense of American liberty, and it was there decided that the decaying institutions of the Middle Ages should not prevail in America, but that just laws and priceless liberty should be planted forever in the domains of the New World.

Historians, becoming engrossed with the more stirring scenes of the Revolution, have failed to consider the sanguinary battle in its true import and bearing upon the destiny of our country, forgetting, that the Colonial army returned home only to enlist in the patriot army and on almost every battlefield of the Revolution were representatives of that little band who stood face to face with the savage allies of Great Britain at Point Pleasant."

And, in conclusion, Kercheval says, at page 139, "I say, from all which it will appear that Dunmore had his views, and those views hostile to the liberties of America, in his proceedings with the Indians in the war of 1774, the circumstances of the times, in connection with his equivocal conduct, leads us almost naturally to infer that he knew pretty well what he was about, and among other things, he knew that a war with the Indians at this time would materially subserve the views and interest of Great Britain, and consequently he perhaps might feel it a duty to promote said war, and if not, why betray such extreme solicitude to single out some conspicuous character, and make him the scape-goat, to bear all the blame of this war, that he and his friend Connolly might escape?"

Nothing could more fittingly describe the patriotic sentiment fell in Virginia than the heroic appeal of Mrs. Wm. Lewis. It is related of her that "When the British force under Tarleton drove the legislature from Charlottesville to Staunton, the stillness of the Sabbath eve was broken in the latter town by the beat of the drum, and volunteers were called for to prevent the passage of the British through the mountains at Kockfish Gap. The elder sons of Wm. Lewis, who then resided at the old fort, were absent with the northern army. Three sons, however, were at home, whose ages were 17, 15 and 13 years. Wm. Lewis was confined to his room by sickness, but his wife, with the firmness of a Roman matron, called them to her, and bade them fly to the defense of their native land. "Go my children.' said she, "I spare not my youngest, my fair-haired boy, the comfort of my declining years. I devote you all to my country. Keep back the foot of the invader from the soil of Augusta, or see my face no more." When this incident was related to Washington, shortly after its occurrence, he enthusiastically exclaimed, "Leave me but a banner to plant upon the mountains of Augusta, and I will rally around me the men who will lift our bleeding country from the dust, and set her free." Howe's *Virginia, Its History and Antiquities*, p. 183.

From Wither's *Border Warfare* we quote: "The army under Gen. Lewis had endured many privations and suffered many hardships. They had encountered a savage enemy in great force, and purchased a victory with the blood of their friends. When arrived near to the goal of their anxious wishes, and with nothing to prevent the accomplishment of the object of the campaign, they received those orders with evident chagrin, and did not obey them without murmuring. Having, at his own request, been introduced severally to the officers of that division, complimenting them for their gallantry and good conduct in the late engagement, and assuring them of his high esteem, Lord Dunmore returned to his Camp; and Gen. Lewis commenced his retreat. "

"This battle (says Col. Stuart, in his historical memoir) was, in fact, the beginning of the revolutionary war, that obtained for our country the liberty and independence enjoyed by the United States and a good presage of future success; for it is well known that the Indians were influenced by the British to commence the war to terrify and confound the people, before they commenced hostilities themselves the following year at Lexington. It was thought by British politicians, that to excite an "Indian war would prevent a combination of the colonies for opposing parliamentary measures to tax the Americans. The blood, therefore, spilt upon this memorable battlefield, will long be remembered by the good people of Virginia and the United States with gratitude."

Virgil A. Lewis, West Virginia State Historian and Archivist, says, in his *History of West Virginia*, published in 1889, at page 133, "To the student of history no truth is more patent than this, that the battle of Point Pleasant was the first in the series of the Revolution, the flames of which were then being kindled by the oppression of the mother country, and the resistance of the same by the feeble but determined colonies. It is a well known fact that emissaries of Great Britain were then inciting the Indians to hostilities against the frontier for the purpose of distracting attention and thus preventing the consummation of the union which was then being formed to resist the tyranny of their armed oppressors. It is also well known that Lord Dunmore was an enemy to the colonists, by his rigid adherence to the royal cause and his efforts to induce the Indians to co-operate with the English, and thus assist in reducing Virginia to subjection. It has been asserted that he intentionally delayed the progress of the left wing of the army that the right might be destroyed at Point Pleasant. Then, at the mouth of the Great Kanawha river, on the 10th day of October, 1774, there went whizzing through the forest the first volley of a struggle for liberty which, in the grandeur and importance of its results, stands without a parallel in the history of the world. On that day the soil on which Point Pleasant now stands drank the first blood, shed in defense of American liberty, and it was there decided that the decaying institutions of the Middle Ages should not prevail in America, but that just laws and priceless liberty should be planted forever in the domains of the New World. Historians, becoming engrossed with the more stirring scenes of the Revolution, have failed to consider this sanguinary battle in its true import and bearing upon the destiny of our country, forgetting that the colonial army returned home only to enlist in the patriot army, and on almost every battle-field of the Revolution represented that little band who stood face to face with the savage allies of Great Britain at Point Pleasant.'

Owing to the importance of the question, we have, at the risk of tiring the reader, given these many details of evidence that the Battle of Point Pleasant, while not a battle between the English and Colonial forces, nevertheless shed the first blood on American soil for national independence. It can be plainly seen that, though at this time these sturdy pioneers were fighting to protect their homes and firesides, the very foundation of national government, Great Britain, through her Tory Governor of Virginia, intended thus to destroy the flower of the Colonial army of Virginia. It was a stroke which, had it succeeded, would have averted the War of the Revolution many years. The army that Lewis gathered were not the unlettered men of the forest, they were from among the most highly educated men of the colony and it is said that, to this date, in no army of a similar number, has such a large percentage had a knowledge of the Greek and Latin languages That they were men of education and influence will be seen by following the survivors of that battle, not only through the Revolution, where many of them distinguished themselves, but out into the civil life of the country, during, and subsequent to, the Revolution.

That the battle was the most fruitful, in its results, of any battle ever fought upon American soil, is apparent from the history of the country. The great Northwest Territory, lying north of the Ohio and east of the Mississippi, had long been a bone of contention between France and England and France did not relinquish her claim until driven to recede as the result of the battle upon the Plains of Abraham before Quebec, where the intrepid Montcalm was defeated by the invincible Wolffe.

The treaty that followed at Paris, in 1763, ceded all this territory to England, whose failure to open it to the colonists was a subject of discussion and distrust and rightfully so, as England maintained it to the exclusion of the colonists, not only that she might, with it, subsidize the savage Indians, but when necessary, secure their services in maintaining control of the colonies.

By the treaty that followed the battle of Point Pleasant, that of Camp Charlotte, the federation of the five great nations in control of that territory ceded it to Virginia, to hold inviolate, and which treaty lasted without interruption for three years, enabling the colonists not only to enter the Northwest Territory, but to colonize Kentucky and Tennessee. In Dunmore's army was the intrepid George Rodgers Clarke, a Virginian, the Hanibal of the West, who was present at the treaty of Camp Charlotte. The history of the colonization and civilization of this territory is the history of Geo. Kcdgers Clarke, too well known here for extended comment. Suffice it to say that, in the struggle led by Clarke to drive the British from the Northwest Territory, it was not the colonies, but Virginia, protecting her own territory, acquired by the battle of Point Pleasant, that furnished the army for Clark's expedition, Governor, Patrick Henry supplying Clarke from Virginia's funds, the sum of twelve hundred pounds, and supplies of boots and ammunition from Pittsburg, then in

Virginia. Could any army have displayed more heroism, an army of one hundred and fifty, starting out to conquer such a wilderness, with no conveyance for their munitions of war, save their own robust and hardy bodies?

The subjugation of this country was not only comparatively broad in its results, but was due alone to Virginia. Of course, such a vast territory opened up, as it thus was, to civilization and habitation, necessarily called for representation in the Congress of the infant nation, and justly so. Virginia would soon, by her great population, control the legislation of the nation. Such, however, was not the purpose of Virginia. That ever generous mother state here had opportunity to be the most magnanimous of them all. She would not, if she could, dominate the policy of the country, and, without a dollar, she donated, actually gave away to the colonies in fee simple the entire Northwest Territory, to be the territory of the colonists, and to be disposed of as they deemed best.

When we review the acquisition of the other territory of the United States and compare the $16,000,000, expended by our government, for the Louisiana purchase, the cost of the acquisition of upper and lower California, of Alaska, of the Philippines, of the cost of the Mexican acquisition in men and money, and then remember that the settlement of the states of Kentucky and Tennessee were made possible, as well as the colonization of Western Pennsylvania and Western Virginia, together with the acquisition of the Northwest Territory, and the settlement and civilization of the same, and all as a sequel of the Battle of Point Pleasant, considering the history of the ever memorable struggle and the subsequent development of the country, it is very apparent not only that the Battle of Point Pleasant was the initial, the first battle of the Revolution, but also farther reaching in its results than any other battle ever fought upon the American continent.

As we have said before, no official report of the battle was ever made, but a letter from Williamsburg, Va., then the seat of government, under date of November 10, 1774, was published in the Belfast News Letter, yet preserved. Presumably, it was contributed to that paper because many of the Scotch-Irish had emigrated from Belfast, Ulster District, to Pennsylvania and ultimately to Virginia and settled in the sections of Virginia from which the army had been for the most part been made up. This made the event peculiarly interesting to the people of that portion of Ireland. From that publication we quote the history of the battle:

"BELFAST.

Yesterday arrived a mail from New York, brought to Falmouth by the Harriot Capt. Lee.

Williamsburg, Va.,

November 10th.

The following letter is just received here from the camp at Point Pleasant, at the mouth of the Great Kenhawa (as then spelled), dated October 17, 1774:

"The following is a true statement of a battle fought at this place on the 10th instant: On Monday morning, about half an hour before sunrise, two of Capt. Russell's company discovered a large party of Indians about a mile from the camp, one of which men was shot down by the Indians; the other made his escape, and brought in the intelligence. In two or three minutes after, two of Capt. Shelby's company came in and confirmed the account.

"Col. Andrew Lewis, being informed thereof immediately ordered out Col. Charles Lewis, to take command of one hundred and fifty of the Augusta troops, and with him went Capt. Dickinson, Capt. Harrison, Capt. Wilson, Capt. John Lewis of Augusta, and Capt. Lockridge, which made the first division. Col. Fleming was also ordered –to take command of one hundred and fifty more of the Botetourt, Bedford and Fincastle troops: Capt. Thomas Buford, from Bedford; Capt. Love, of Botetourt; Capt. Shelby and Capt. Russell, of Fincastle, which made the second division.

"Col. Charles Lewis' division marched to the right, some distance from the Ohio, and Col. Fleming, with his division on the bank of the Ohio, to the left.

"Col. Charles Lewis' division had not marched quite half a mile from the camp when, about sunrise, an attack was made on the front of his division, in a most vigorous manner, by the united tribes of Indians Shawnees, Delawares, Mingoes, Tawas, and of several other nations in number not less than eight hundred, and by many thought to be one thousand.

"In this heavy attack, Col. Charles Lewis received a wound which, in a few hours caused his death, and several of his men fell on the spot; in fact, Augusta division was obliged to give way to the heavy fire of the enemy. In about a second of a minute after the attack on Col. Lewis' division, the enemy engaged the front of Col. Fleming's division, on the Ohio, and in a short time the Colonel received two balls through his left arm, and one through his breast, and, after animating the officers and soldiers in a most calm manner to the pursuit of victory, retired to the camp.

"The loss in the field was sensibly felt by the officers in particular; but the Augusta troops, being shortly after reinforced from the camp by Col. Field, with his company, together with Capt. McDowell, Capt. Mathews and Capt. Stewart, from Augusta; Capt. Paulin, Capt. Arbuckle and Capt. McClannahan, from Botetourt, the enemy no longer able to maintain their ground, was forced to give way till they were in a line with the troops, Col. Fleming being left in action on the bank of the Ohio.

"In this precipitate retreat Col. Field was killed. During this time, which was till after twelve, the action in a small degree abated, but continued, except at short intervals, sharp enough till after 1 o'clock. Their long retreat gave them a most advantageous spot of ground, from whence it appeared to the officers so difficult to dislodge them that it was thought most advisable to stand as the line was then formed, which was about a mile and a quarter in length, and had sustained till then a constant and equal weight of the action, from wing to wing. "It was till about half an hour till sunset they continued firing on us scattering shots, which we returned to their disadvantage. At length, the night coming on, they found a safe retreat.

“They had not the satisfaction of carrying off any of our men's scalps, save one or two stragglers whom they killed before the engagement. Many of their dead they scalped, rather than we should have them, but our troops scalped upwards of twenty of their men that were first killed.

"It is beyond doubt their loss; in number, far exceeded ours, which is considerable.

"The return of the killed and wounded in the above battle, same as our last, as follows:

Killed - Colonels Charles Lewis and John Field, Captains John Murray, R. McClannahan, Samuel Wilson, James Ward, Lieut. Hugh Allen, ensigns Cantiff and Bracken, and forty-four privates. Total killed, fifty-three.

"Wounded - Col. William Fleming, Captains John Dickinson, Thomas Buford and I. Skidman, Lieutenants Goldman, Robinson, Lard and Vance, and seventy-nine privates. Total wounded, eighty-seven; killed and wounded one hundred and forty."

And further from the same publication:

"AMERICA.

Williamsburg, in Virginia, December 1, 1774.

We have it from good authority that his excellency, the governor, is on his way to this capital, having concluded a peace with the several tribes of Indians that have been at war with us, and taken hostages of them for their faithful complying with terms of it, the principal of which are that they shall totally abandon the lands on this side of the Ohio river, which, river is to be the boundary between them and the white people, and never more take up the hatchet against the English."

"Thus, in a little more than the space of five months, an end is put to a war which portended much trouble and mischief to the inhabitants on the frontier, owing to the zeal and good conduct of the officers and commanders who went out in their country's defense and the bravery and perseverance of all the troops.' Copied from the *Belfast News Letter* of February 10, 1775."

De Hass, in describing the battle, says:

"The battle scene was now terribly grand. There stood the combatants terror, rage, disappointment and despair riveted upon the painted faces of one, while calm resolution and the unbending will to do or die were marked upon the other. Neither party would retreat, neither could advance. The noise of the firing was tremendous: no single gun could be distinguished was one common roar. The rifle and the tomahawk now did their work with dreadful certainty. The confusion and perturbation of the camp had now arrived at its greatest height. The confused sounds and wild uproar of the battle added greatly to the terror of the scene. The shouting of the whites, the continued roar of firearms, the war-whoops and dismal yelling of the Indians, were discordant and terrific."

Col. J. L. Peyton, in his valuable history of Augusta county, says:

"It was, throughout, a terrible scene the ring of rifles and the roar of muskets, the clubbed guns, the flashing knives the fight, hand to hand the scream for mercy, smothered i n the death-groan the crushing through the brush the advance the retreat the pursuit, every man for himself, with his enemy in view the scattering on every side the sounds of battle, dying away into a pistol shot here and there through the wood, and a shriek the collecting again of the whites, covered with gore and sweat, bearing trophies of the slain, their dripping knives in one hand, and rifle-barrel, bent and smeared with brains and hair, in the other. No language can adequately describe it."

Mr. Stephen T. Mitchell in 1827 in a publication, *The Spirit of the Old Dominion* published at Richmond Virginia gives the following account of the battle of Point Pleasant.

"We landed about a mile on the left-hand shore of Kanawha, and climbing a large hill, we were saluted by a hundred Indians, encamped upon the top. Our captors told their adventures, no doubt, with every aggravation; for, after the most frantic expressions of grief and rage, I was bound to a tree, a large pine tree, which stands to this day upon the brow of the hill, and the fire was kindled around me. I said my prayers; my time was come; my body felt the scorching heat: but, by a miraculous interposition of Providence, the clouds which had been lowering all day, now burst out in showers, and quenched the flames. The Indians thought the Great Spirit looked over me, and directed the shower for my safety. My bonds were loosened, and I was allowed a little jirk and hommony for my refreshment. The next day I could perceive some great expedition on foot; the Indians were running to and fro in every direction; some grinding paint and some cleaning up their arms; and even the squaws and little boys were providing themselves with hatchets and scalping-knives, and strewing themselves from the Ohio river all along the cliffs of Kanawha."

"Late in the evening, I saw an uncommon anxiety on the faces of the savages; councils, grand and petty, were held in various places, and so completely were my guards absorbed in the undertaking which was at hand, that they became entirely remiss in their attentions to me. I resolved to seize the propitious moment, and make my escape. I sprang on my feet and ran as fast as my legs would carry me. A loud whoop proclaimed the event, and in a moment, I could perceive myself closely pursued by half a dozen athletic young fellows, with uplifted tomahawks. Fear added to my limbs the agility of the deer. With my head turned back over one shoulder, I bounded through the pine-trees until my speed had carried me unawares to the brink of a precipice. I tried to stop; it was too late; I gave a piercing shriek and bounded over. A rushing sound in my ears like the roaring of a mill-dam, then the crashing of branches and limbs recalled me to my recollection, and I found myself to my inexpressible delight, breaking my way through the thick branches of a buck-eye tree. I alighted without injury, and looking back upon the cliff above, could see my savage pursuers gaping over the precipice in amazement. I gave not a second look, but darted off towards the point with a heart swelling with praise to the great Creator, who had thus twice rescued me so miraculously from my enemies. Arriving at the mouth of the Kanawha, I shouted aloud for assistance. But, the whites had too often been decoyed by their own people to the savages, to be easily imposed upon. They answered me they could give no assistance. I could not swim, but my ingenuity, never fertile in expedients, befriended me now tor the first time in my life. I rolled down a dry log from the bank into the water, and getting astride of it, I managed by great exertion of hands and feet, to row it across the stream, which at that time, from the great height of the Ohio, was as still as a mill-pond I was received by General Lewis, the commandant of the fort, with great cordiality and affection; and, being naked and necessitous, I enrolled myself as a regular in the corps; and, being dressed in militaire, with a tremendous rifle in my hand and a thick breast work before me, I felt as brave as Julius Ceasar."

"I was in hopes that I might enjoy, within the walls of a fort, some respite from the fears, toils and anexieties which had, for the last two weeks, worn me out both body and mind. But he who undertakes to settle in a new and savage country, must look out for no such respite, until, by hardihood and perseverance, he has levelled the forest, with its inhabitants, to the earth.

On the 10th of October, 1774, about sunrise, the hunters came in at fall speed, and gave the appalling information that a large body, of Indians had spread themselves from river to river, and were advancing by slow degrees, towards the fort; at the same instant, we could observe the women and boys skulking up and down the opposite banks of the Ohio and Kanawha.

The position of the fort was peculiarly favourable to a surprise. As I have above mentioned, it was situated at a right angular point formed by the confluence of the Kanawha and Ohio rivers. The country above the fort was covered with a heavy forest and impervious growth of underwood, through which an invading force might penetrate completely undiscovered, to the very walls of the fort. The garrison was composed of about twelve hundred men entirely Virginians, from the counties of Botetourt and Augusta. The Indians consisted of about the same number, the flower of the Shawnee, Wyandotte and Mingoe tribes, who were commanded by the celebrated Chieftain, Cornstalk."

"From the large force which he had collected for this expedition, and from the secrecy of his movements, it was evident that the Indian Chief, in this desperate attempt to recover the country east of the Ohio river, meditated nothing less than an entire extermination of the garrison. General Lewis ordered out about seven hundred of his rangers, under the command of his nephew, Colonel Charles Lewis; with the remaining part of his troops, about five hundred in number, he determined to act as a reserve and defend the fort to extremities. "

"I happened to be among those who were ordered out, very much against my will; but it was neck or nothing; we advanced about three hundred yards in front of the fort, toward a deep ravine which intersected the valley at the right angles with the Kanawha. All was still as death; one moment more and a yell mingled with the roar of a thousand rifles, rung from river to river, and at the same moment every bush and tree seemed alive with armed savages. Col. Lewis was killed at .the first fire, but the rangers maintained their ground, and a contest commenced more desperate and more rapidly fatal than any which had ever been fought with the Aboriginese, excepting that of Talledaga. The Indian Chief, with that promptness for seizing an advantage, and that peculiar military tact for which he was so much renowned, extended his line from the Ohio as far as it would stretch across to the Kanawha bank, for the purpose of outflanking the opposing forces. But, in the execution of this manoeuvre, he was completely foiled by the superior address and boldness of the whites who, animated with revenge for the loss of their leader and a consciousness of their desperate situation, fought with a fury that supplied the inequality of numbers, and set at defiance every stratagem of the savages."

"Finding that his method of outflanking would not succeed, the Indian Chief concentrated his forces, and furiously attacked the centre of the Virginia line. The savages, animated by their warlike and noble Chieftan, Cornstalk, forgot the craftiness of their nature, and rushing from their coverts, engaged hand to hand with their stout and hardy adversaries, until the contest resembled more a circus of gladiators than a field of battle. I became desperate; hide where I would, the muzzle of some rifle was gaping in my face, and the wild, distorted countenance of a savage, render he more frightful by paint, was trushing towards me with uplifted tomahawk One fellow in particular, seemed to mark me as his victim; I levelled my rifle at him as he came yelling and leaping towards me, and fired. The ball missed my aim. He rose upon his toes with exultation, and whirling his tomahawk round his head, slung it at me with all his powers. I fell upon my face, and it whizzed harmless over my head and stuck into a sapling. I bounded up and forced it from the tree, but the Indian was on me and rescued the hatchet from my hands. I seized him round the waist, enclosing both his arms at the same time and tripping up his heels, we rolled together upon the ground. I at last grew furious, gouged him with my thumbs in both eyes, and seizing him with my teeth by the nose, I bit the whole of it from his face; he yelled out with pain and rage, and letting loose the hatchet to disengage my teeth, I grasped the handle and buried the sharp point into his brains. He gave one convulsive leap which bounced me from his body, and in a moment after expired. I immediately rose, and gaining a secure position behind a tree, remained there till the close of the fight, and made a thousand resolutions, if I survived this engagement, never to be caught in such a scrape again. I kept my word; for, I have never since encountered the savages, and if Heaven forgives me, I never will. There is no fun in it."

"But, to return to the history of this ever memorable battle. There was a peninsula extending from a high range running parallel with the Ohio river, which jutted close to the Kanawha bank, about a half a mile from its mouth. Knowing the importance of securing the narrow pass which ran between its base and the river, the Indian Chief despatched a picked body of his troops to take possession of it. They entered the dry bed of a small creek which skirted the foot of the hills, and pursued their route unnoticed till they were about to enter the important pass, when a shower of rifle bullets pierced their body and swept down the foremost ranks. A chosen band of rangers at the same moment made their appearance, with whom General Lewis in anticipation had guarded the pass. A yell of surprise and rage burst from the savage line, and they seconded their returning fire by an unanimous and desperate charge with the hunting knife. The contest now assumed all the wild and terrific cast which a personal struggle, conducted with the deadly feelings of hate and revenge then existing between the whites and Indians, could inspire. The air was filled with the screams of the savages and the deep imprecations of the riflemen; every blow brought death, and the ground was soon heaped with the corpses of the combatants. But the disappointed efforts of savage desperation were ineffectual against the unbroken and impenetrable column which was maintained by the whites; and the Indians were driven, with the loss of half their force, back upon the main body. Here, the fight still raged in the extremity of opposition, every inch of ground was contested, from behind every bush and decayed, log the murderous flash arose, and the continued roar of a thousand rifles vibrated through the forest."

"The savage Chieftain discovered that the chances against him were desperate, yet, by his own personal example of courage and address, was the fight long sustained, even after his line had been driven, step by step, from their original position. His voice could at intervals be heard, rising above the din of the fight like the shrill blast of a bugle; at one moment, his dusky form and glittering ornaments could be seen flitting through the trees upon the Ohio bank, and his war cry in the next would fill the echoes of the hill at the farthest extremity of the line. A cheering ejaculation of triumph would one moment escape him, as an advantage was gained by the devoted gallantry of some Shawnee warrior; an imprecation upon some skulking Mingoe, in a short time afterwards, would be recognized in his voice. "Charge high and aim low" was his command incessantly throughout the day; and, it is one of the circumstances remarked of that fatal fight, that most of the bullet-wounds received by the whites proved mortal; but few of the wounded ever recovered. Yet, all the efforts of the old warrior were vain; defeated and discouraged, the savage army almost abandoned the fight in the latter part of the day, and it was reduced to a mere straggling fire between individuals of the contending parties."

"Night closed upon the scene, yet the ground was still occupied by the two armies. Although victorious, the Virginians could neither press their advantage nor retire to rest. An ambuscade or a night attack was expected from the savages, and their behaviour warranted the latter supposition.

For, behind a long line of watch-fires, they could be discovered as if cautiously examining the points most open to attack. The wild scream of a savage warrior, apparently advancing to the fight, would at intervals break upon the death like stillness of the night, and cause my heart to leap almost out of my mouth. I confidently calculated that every moment was the time for their attack, and fancied divers times could hear them stealing through the bushes upon us. The gleams of the morning sun, however, at length illumined the scene, but not a vestige of the Indian army remained; the living and the dead had alike disappeared, and it was not until then, it was ascertained or even suspected, that the savages had secure themselves from interruption, under pretense of a night attack, had thrown their dead, with weights attached to them, in the river, and retreated across it under cover of darkness."

Of the men who participated in the Battle of Point Pleasant, we regret that no complete roster has been preserved. However, the men who were in that army were friends and neighbors, and many of them related by ties of blood and marriage, so that a review of a few of them will indicate (he character of the men composing the army.

It will be seen by a review of the history of the colonies that prior to the Battle of Point Pleasant, not only the Colonists but England knew, as did Patrick Henry when he made his famous speech that "The War was inevitable." The British Government seeing the fomentation in the colonies had made repeated concessions; willing to relinquish, if necessary, all but the principle of the Right of England to levy taxes upon the Colonists without giving them representation in the British Government. The Colonists were astir with intense excitement. The tea had been thrown over board in Boston Harbor and the Port had been closed by a bill passed by Parliament in March of that year. Meetings had been and were being held protesting against Royal oppression. That powerful engine of resistance, Committees of Correspondence had been formulating their ideas of resistance and the Virginia Assembly convened at Williamsburg in May, had passed an independent resolution setting forth that June 1st, 1774, should upon the making effective of the Port Bill be made "a day of fasting and prayer to implore the divine interposition for avering the heavy calamity, which threatens the civil right of America;" whereupon, the Earl of Dunmore, then Governor of Virginia, at once dissolved the Assembly. The Continental Congress had already convened and its every breath was ladened with resistance of British oppression.

Is it to be wondered at and is it not the most natural thing in the world, that Dunmore would try to devise ways and means to prevent Virginia from participating in the federation of the Colonies; and what more powerful instrument could he" have set in motion to distract their attention from the clouds gathering in the East, than by setting in motion a band of howling Indians on the frontier, making it an absolute necessity that Virginia protect her homes, her women and children and her property rights, and this danger so eminent, could not be delayed. So calling together the flower of the Colonial Army of Virginia, which he promised should be united and together encounter the Indians in their homes, he should cause one branch to alone be attacked, hoping they would thus be destroyed and if only temporarily defeated, they would be so busy protecting the frontier and their homes they would have no time to go into the Colonial Army, confederated as they would be to resist the British Army, already many of whom were camping upon the plains of Boston. But to the surprise of Dunmore the Division of Lewis' Army was victorious and the tide of American interests was changed.

Without the Army of Lewis, which was the great military training school of the Colony, many of whom went on into the Revolution and became many of them, officers of high rank, it would have been impossible for Virginia to have raised her quota of men and officers to have participated in that struggle for liberty; and without Virginia the Colonists would have thought it impossible, as it would have been, to have undertaken that struggle for independence. Without the entire support that Virginia gave George Rodgers Clark who was in the Dunmore division, but who later conquered the North West Territory, weakening the otherwise impregnable background that constantly threatened the frontier and in whose territory did not close the struggle for American Independence until Waynes treaty twenty years later.

We think the opinions of the early writers of history we have quoted, the natural circumstances surrounding Dunmore at and previous to the Battle, makes it plain that although the battle was between the Colonists and Indians it is beyond doubt the first Battle of the Revolution, and the Government of the United States, while it has been tardy, is fully justified in making the declaration that the $10.000 appropriated for the erection of a monument is

"An act to aid in the erection of a memorial structure at Point Pleasant, West Virginia, to commemorate the Battle of the Revolution, fought at that point between the Colonial troops and Indians, October 10th, seventeen hundred and seventy four."

While a shaft 82 feet high will stand as a sentinel upon the site where the dead were buried, from whence the battle was directed and subsequently the fort, built, it is a pigmy as compared with the fact that at last, after a lapse of One Hundred and thirty-four years, the Congress of the United States has officially called it as it is a battle of the Revolution, and if a battle of the Revolution it must of necessity be the first, as the hallowed Lexington was not fought, until April 19th, 1775, while that of Point Pleasant, was fought October 10th, 1774.

The battle in its acquisition of territory ceded by the Indians and previously ceded by France to Virginia but literally in control of the Indians until this time, this followed by the ceding of all the vast territory of the Great North West by Virginia to the infant republic at the close of the Revolution with the cessation of Indian hostilities following the battle, permitting the Colonists to turn their attention to the expulsion of the English army and the overthrow of the British yoke, the moral effect that it had on Virginia, and thus on the Colonies, made it the farthest reaching in its effect an battle ever fought on the American Continent.

The name of every man who participated in that struggle whether he protected the frontier nearer home while the band of stalwarts went forth to conquer the Indians and make secure the wilderness, the men of Wm. Christian's Regiment who rendered such valiant service, coming as they did when the battle was over, the army exhausted wounded and bleeding and in time to gather up and bury the slain, should all be honored and preserved. Christain's men were only delayed by their effort to bring in supplies to the Army of 54000 pounds of flour on 400 pack horses but 108 additional head of cattle. They expected to join Lewis' Army and together march on to encounter the Indians upon the Pickaway Planes; so that as a part of the Army they are entitled to be enrolled with the heroes of that battle, which will be followed by the roster so far as the writer has been able to glean from all avail- able sources, after many years of careful research.

BIOGRAPHICAL

GENERAL ANDREW LEWIS

GENERAL ANDREW LEWIS

General Andrew Lewis Gen. Andrew Lewis, the hero of the Battle was not only a gentleman of education and refinement, but was a past master in the art of military tactics, Leaving entered upon his career in 1742 as Captain of Militia and ten years later as head of Militia of his county. He was with Washington at Great Meadows and Fort Necessity in 1754, when he was twice wounded. In 1755, he was detailed to build forts hence was not present at Braddock's disaster. In 1756, he led the Big Sandy expedition against the Shawnees. In 1758, when, with Washington and Gen. Forbes, at Fort DuQuesne, he was wounded.

He surrendered to a French officer, was imprisoned at Montreal, was exchanged and saw active service. In 1762 at his request his company was disbanded. In 1763 he was appointed to Lieut. for Augusta Co.

We next find him a member of the Virginia House of Burgesses and a member of a committee to negotiate treaties. It was while thus engaged at the treaty of Fort Stanwix that the Governor of New York said of him, "He looks like a genius of the forest, and, the very ground seems to tremble under him as he walks along." It is from Col. Stewart, his biographer, we learn that "He was upwards of six feet high, of uncommon strength and agility, and his form of the most exact symmetry He had a stern countenance and was of a reserved and distant deportment, which rendered his presence more awful than engaging." While Mr. Alexander Reed, of Rockbridge County, Virginia, who was with him at Point Pleasant, describes him thus; "He was a man of reserved manners, and great dignity of character somewhat of the order of General Washington." General Washington held him in such esteem that ne recommended that he be made commander-in-chief of the Continental army.

The Battle of Point Pleasant was not only the pivotal point in the life of the nation, but in the life of General Lewis as well. Heretofore he had fought as a British subject. In defying the orders of Lord Dunmore, the Tory Governor, he was not only among the first to defy Briton, but the first to take up arms in defiance of British authority and from this time on we find him enlisted in the cause of the colonists as against the English Crown. It seems the irony of fate that he should not have lived to witness the surrender at Yorktown. While enroute home he died of a fever at the home of his friend, Col. Buford, and was taken to his own estate "Richfield" where his remains were interred Sept. 20, 1731, near the town of Salem, where they reposed for many years in an unmarked and neglected grave. A few years ago, the Ladies of the Margaret Lynn Lewis Chapter, Daughters of the American Revolution, of Roanoke, Virginia, had his remains (which were found well preserved) removed to the East Hill Salem Cemetery, where, in 1902, they erected a stately shaft to his memory. One of the six figures of Roger's equestrian statute at Richmond, commemorating famous Virginians, is that of General Andrew Lewis, but the greatest of his monuments is builded in the hearts of a grateful American Republic.

Beside his brother Col. Charles Lewis, and John, son of his brother William, Andrew had three sons in the Battle of Point Pleasant. What greater love can a man have for his country than, like Jacob of old, to lay his sons as a sacrifice on his country's altar?

COL. CHARLES LEWIS

While Andrew Lewis is known as the "Hero of the Battle of Point Pleasant," his brother, Col. Charles Lewis, a brave soldier, too, was called "The idol of the army." While Andrew had devoted his life to the cause of his adopted country, he having been born in Ireland where his parents were then residing, it was reserved for Col. Charles to embody the completeness of American association, he having been born in America, being the youngest child of John Lewis and Margaret Lynn Lewis. Thus he had the distinction not only of dying on American soil, but also of having been born there; in 1733, in the county of Augusta, State of Virginia, and was thus all his life known of Virginia, loved of Virginia and he sacrificed his life, satisfied that he had given to Virginia her lull measure of devotion. He was mortally wounded while leading a division of the army at the outset of the Battle of Point Pleasant and later was led to his tent where in a few hours he expired. Col. Andrew Lewis, his nephew who was engaged in the battle, says "He received his wound early in the action but did not let it be known until he had gotten the line of battle extended from the Ohio to Crooked Creek, after which he asked Captain Murray, his brother in law, to let him lean on his shoulder and walk with him to his tent, where he expired about 12 o'clock." Captain Arbuckle states that he received a wound which in a few hours caused his death. Roosevelt's winning of the West says "The attack fell first and with especial fury, on the division of Charles Lewis who himself was mortally wounded at the very outset, be had not taken a tree (the frontier expression for covering oneself behind a tree trunk) but was in an open piece of ground, cheering on his men when he was shot. He stayed with them until the line was formed, and then walked back to camp unassisted, giving his gun to a man who was near him." Howe says of him. "Charles Lewis was esteemed the most skillful of all the leaders of the border warfare and was as much beloved for his noble and amiable qualities as he was admired for his military talents." On page 182 of Howe's *Virginia Its History & Antiquities*, we find a sketch from his life;

"Charles Lewis, the youngest son of John, is said never to have spent one month at a time out of active and arduous service. Charles was the hero of many a gallant exploit, which is still treasured in the memories of the descendants of the border riflemen, and there are few families among the Alleghanies where the name and deeds of Charles Lewis are not familiar as household words. On one occasion, Charles was captured by the Indians while on a hunting excursion, and after having traveled some two hundred miles, barefoot, his arms pinioned behind him, goaded on by the knives of his remorseless captors, he effected his escape. While traveling along the bank of a precipice some twenty feet in height, he suddenly, by a strong muscular exertion, burst the cords which bound him, and plunged down the steep into the bed of a mountain torrent. His persecutors hesitated not to follow. In a race of several hundred yards, Lewis had gained some few yards upon his pursuers, when, upon leaping a prostrate tree which lay across his course, his strength suddenly failed, and he fell prostrate among the weeds which had grown up in great luxuriance around the body of the tree. Three of the Indians sprang over the tree within a few feet of where their prey lay concealed; but with a feeling of the most devout thankfulness to a kind and superintending Providence, he saw them one by one disappear in the dark recesses of the forest. He now bethought himself of rising from his uneasy bed, when lo a new enemy appeared, in the shape of an enormous rattlesnake, who had thrown himself into the deadly coil so near his face that his fangs were within a few inches of his nose; and his enormous rattle, as it waved to and fro, once rested upon his ear. A single contraction of the eyelid a convulsive shudder the relaxation of a single muscle, and the deadly beast would have sprung upon him. In this situation he lay for several minutes, when the reptile, probably supposing him to be dead, crawled over his body and moved slowly away. "I had eaten nothing,' said Lewis to his companions, after his return, "for many days; I had no fire-arms, and I ran the risk of dying with hunger, ere I could reach the settlement; but rather would I have died, than made a meal of the generous beast'."

Kercheval's *History of the Valley*, describes the attire of Col. Charles Lewis on that day, at page 114, as follows: "Col. Chas. Lewis, who had arrayed himself in a gorgeous scarlet waistcoat, against the advice of his friends, thus rendering himself a conspicuous mark for the Indians, was mortally wounded early in action; yet was able to walk back after receiving the wound, into his own tent, where he expired. He was met on his way by the commander-in-chief, his brother, Col. Andrew Lewis, who remarked to him," I expected something fatal would befall you," to which the wounded officer calmly replied, "It is the fate of war." The same author says at page 115, "Col. Lewis, a distinguished and meritorious officer, was mortally wounded by the first fire of the Indians, but walked into the camp and expired in his own tent."

Peyton's *History of Augusta County* says "He abandoned himself too much to his passion for glory and forgot the wide difference between an officer and a private, fie was not inferior to his brother, the General, in courage, intrepidity and military genius; he surpassed him in some respects, he knew how to oblige with a better grace, how to win the hearts of those about him, with a more engaging behavior. He consequently acquired the esteem and affection of his men, in a most remarkable manner. To perpetuate the memory of his public and private virtues, his eminent services in the field and his heroic fate, the General Assembly of Virginia, in 1816, named Lewis County in his honor."

By his kinsmen be was considered the "flower of the flock." Like his brother, he was a man of splendid physique and without disparaging his kindred, he was best loved because of his high degree of morality, spotless integrity and acknowledged bravery. His long and active military career had made him a hero in the eyes of his comrades from the Braddock campaign to the hour of his death, and while had he lived, he doubtless would have added new lustre to his name in the continued struggle of the Revolution; after all, he had but one life to offer up to his country and at this crucial moment no doubt it was needed most. His conduct inspired the army. The sacrifice of his life armed anew his companies and stimulated them to greater feats of daring. Thwaite's *Dunmore War* says "Charles Lewis was popular and beloved by all the western army. His loss was a general affliction." Dr. Bale's *Trans-Allegheny* says: "Colonel Charles Lewis was said to be the idol of the army. He had a large, active and honorable military experience from Braddock's War down to death. And it is believed that he would have achieved greater honors and distinction in the Revolutionary struggle, if his life had been spared, but his brilliant career was ended in glory on this field."

The charge he made at Point Pleasant was in the face of a fearless band of adversaries. When Gen. Andrew Lewis selected his brother to take command of the left wing of the army in the first attempt to repulse Cornstalk and his fearful braves, he selected his brother to bear that peril, not that he loved him less, but that he knew the army needed his courageous example. Col. William Preston, in writing of his death to Patrick Henry, said: "Poor Charles Lewis was shot in clear ground, encouraging his men to advance. If the loss of a good man, a sincere friend, a brave officer deserves a tear he certainly is entitled to it."

At the close of the conflict, his mortal remains were laid to rest upon the reservation of forty feet square upon the present sight of Tu Endie Wei Park, where the Kanawha and Ohio meet. Here, he is buried beside the other dead of that battle.

No stone as yet has ever marked his resting place, save the four granite corner stones erected in

1905 by the Col. Charles Lewis Chapter Daughters of the American Revolution at Point Pleasant. While no monument has as yet been reared to mark the last resting place of this great man, a tribute due him from his own loved ones, as well as from a grateful nation; it is equally a matter of congratulation that though tardy the government has donated a small amount 10.000 which with the $6.000 in the bands of the State Commission has been pledged in the contract let for a monument. But greater than this monument is the recognition of the Government of the status of the battle as regards the Revolution, standing, as it does, on the heels of Indian depredations on the western frontier and on the threshold of the American Revolution for American Independence. This honor so long-delayed, will at last have written this page of American history correctly when a stately monument shall bear the inscription:

Battle of Point Pleasant

October 10, 1774.

First Battle of the American Revolution.

COL. JOHN FIELD

Col. John Field born in Culpepper County in 1720 of good family is buried beside Col. Charles Lewis, his friend and comrade. He received his fatal wound at the Battle of Point Pleasant, while bravely checking the Indians. He rallied his troops who had begun to waver, and, for a time, seemed likely to be repulsed. Col. Field had been with Braddock and had seen frontier service from that date, both as a militia officer and as a surveyor. In 1765 he served in the Virginia Legislature, in 1766 he was made Colonel of Militia. His troops at Point Pleasant were a volunteer company, raised by him in his own county which he united with that of Gen. Lewis.

COL. WILLIAM FLEMMING

Col. William Flemming was renewed not only as a military genius, but as a learned physician and gentleman of culture. While twice severely wounded at the Battle of Point Pleasant, he recovered and was subsequently acting Governor of the State of Virginia. He was a Scotchman of proud lineage.

Born Feb. 18, 1729. Prior to Dunmore's war he was Lieut. under Gen. Forbes in 1760-61. In 1762 he was Capt. under Col. Adam Stephens at Vaux's and Stonakers forts. In 1763 he married Anne Christian. His home was at "Belmont" in Montgomery Co. The wounds he received at Point Pleasant disabled him for active participation at the war of the Revolution, but be was County Lieutenant, in further defense, in his country against Indians and State Senator 1780-81, acting Governor during the Cornwallis invasion 1781 Twice commissioner to settle local battles with Kentucky; member Virginia convention ratifying the constitution of 1788. He was benevolent and beloved and as a physician and surgeon his ministration to humanity was most extensive. His death Aug. 24th, 1795 was the result of wounds received at the Battle of Point Pleasant.

CAPT. EVAN SHELBY

Capt. Shelby, who with his two sons, was in the Battle of Point Pleasant, was a ranking officer after the death of Colonels Lewis and Field and the wounding of Fleming, until the arrival of Colonel Christian's regiment. It was Evan Shelby who, with his and the companies of Mathews and Stewart in the ruse, executed the flank movement up Crooked Creek, that led Cornstalk to believe that Lewis had been re- enforced, possibly by Dunmore who had played him false and thus decided the fortunes of the day. Captain Shelby became Colonel Shelby of the Revolution, whose distinguished career is followed in the history of Kentucky and Tennessee.

He was born in Wales in 1720, He early emigrated to Maryland. He commanded a company under Forbes in 1758. In 1771 he settled on the Holston. In 1776 he was with Christian in the Cherokee expedition. In 1779 he led a successful expedition upon the Chickamauga towns, having been appointed by Virginia a general of Militia. He continued in the service throughout the Revolution engaged principally upon the sea board. He rose to the rank of Colonel and before the close of the war to that of General. He died at King's Mountain, Dec., 4th, 1794.

ISAAC SHELBY

Isaac Shelby who served as a Lieutenant under his father Capt. Evan Shelby at the battle of Point Pleasant, was born at North Mountain, Md., Dec., 11th, 1750, where his grand-father settled upon his arrival from Wales. He had removed with his father to the present site of Bristol, Tenn., prior to the Dunmore War and was engaged in feeding and herding cattle. He served throughout the Revolution distinguishing himself at

Camden, King's Mountain and Long Island Flats. Governor Patrick Henry promoted him to a Captaincy and commissioned him Commissary general of the Virginia forces. When Sevier in 1779 projected the expedition that captured the British stores, at Chickamauga, Shelby equipped and supplied the troops by the pledge of his individual credit. Governor Thos. Jefferson commissioned him a Major of Virginia, but a survey proved him to be a resident of North Carolina, when he was appointed a Colonel by Gov. Caswell. He distinguished himself at Thicketty Fork, Cedar Springs and Musgraves Mills. Retreating across the Alleghenies after the disastrous defeat at Camden, he with John Sevier planned the remarkable campaign which resulted in the battle of King's Mountain, the high water mark of the Revolution that turned the tide in favor of the patriot army. He did valiant service at the battle of the Cowpens as well as at Charleston. He was largely instrumental in preserving Kentucky to the Colonists as against an alliance with Spain. He was six times chosen a Presidential elector for Kentucky. In 1812 he became the first Governor of Kentucky, which he accepted with great reluctance and accepted only that he might again aid his country as against Great Britain. He organized 4000 volunteers and at the age of 63 years led them in person to the re-enforcement of Gen'l Wm. Henry Harrison enabling him to profit by the victory of Perry at Lake Erie. Congress voted him a gold Medal, and the Kentucky Legislature a vote of thanks.

In 1783 he married Miss Susan- nah, daughter of Captain Nathanial Hart. He established himself on the first settlement and pre-emption granted in Kentucky which he made his home, residing thereon 43 years. He died July 18th, 1826, aged 76 years. He was a strict Presbyterian.

JOHN JONES

John Jones was born in Culpepper County, Virginia, in 1755, and enlisted in the army of Andrew Lewis and was in the Battle of Point Pleasant, following which, upon his return to Culpepper} he enlisted in the patriot army and served throughout the Revolution. In 1792, he obtained patents for land for 359 acres on the Kanawha River and that same year for 400 acres more. In 1797 he obtained patents for 400 acres in Teays Valley and land on Paint Creek, besides making purchases from the state. He owned from Paint Creek to the Narrows on Kanawha, including the present sites of Pratt and Dego. In the Clifton Cemetery above Paint Creek, his remains were interred, a slab bearing inscription,

"In Memory of John Jones who departed
This life January 7, 1838, Aged 83 Years."

John Jones married Frances Morris, a sister of Win. And Leonard Morris. He was hospitable and a good Baptist Church founded at Kelley's Creek in 1796 was largely due to his interest and generosity.

His will, recorded March, 1838, mentions his wife, Frances, and children, Gabriel, who returned to Culpepper County, Va.; William; Nancy who married --- Huddleston; Thomas; Levi; and Frances, who married Shelton and were the progenitors of the prominent Nicholas County, W. Va., family of that name.

JOHN DRAPER

Lieut. John Draper, of the Battle of Point Pleasant, was born in 1730, and was one of the pioneer settlers of Drapers Meadows, where, in 1755, occurred that dreadful massacre, in which his wife and sister were captured by the Indians. In 1765, he removed to Drapers Valley on the line between Pulaski and Wythe Counties, Va. He was commissioned a Lieutenant in one of the Fincastle Companies in 1774. He died in 1828 at the age of nearly ninety-four years.

Prominent members of the family are still resident of Drapers Valley. In 1886, John S. Draper, a great grandson of Lieut. John Draper, was the owner of the beautiful estate. John Draper was twice married; his first wife was Bettie ---. After her return from captivity, she bore four sons and three daughters, she died in 1774, aged 42 years, and in 1776, he married Mrs. Jane Crockett, who bore him two daughters, Alice and Rhoda. By the first marriage, the sons were George, James, John and Silas. The names of the other two daughters are to us unknown.

BENJAMIN LOGAN

Benjamin Logan, born in Augusta County, Va., 1743, was, in 1764, a sergeant in Boquets's Army. In 1771, he moved to the Holston Valley. He was a Lieutenant in the Point Pleasant Campaign. In 1775, he moved to Kentucky and built Logan's Station or fort which was besieged by Indians. Logan went to Holston settlement for ammunition, returning on foot in ten days. In 1779, he was second in command of the Bowman expedition.

He was a noted Indian fighter and allied himself with Kentucky, as against the Spaniards at New Orleans. He removed to Shelby County, Kentucky, where he died in 1802.

WILLIAM CAMPBELL

William Campbell was born in Augusta County, Virginia, in 1745, died at Rocky Mills, Hanover Co., Va., August 22, 1781. In 1767 he moved to the Holston Valley. In 1774 he was appointed Captain of militia and was in Col. Christian's regiment at Point Pleasant. In September, 1775, he commanded a company at Williamsburg, in Patrick Henry's regiment and under General Lewis assisted in dislodging Governor Dunmore from Gwynn's Island in July, 1776. In 1777 he was made Lieutenant of Militia in the new county of Washington. In 1779, he aided in driving the Tories from the Holston Valley. In 1780, he was promoted to a colonelcy of the regiment and chosen to represent his county in the legislature. At King's Mountain he distinguished himself and was commended by Washington, Gates and Greene. He was with Gen. Greene at Guilford Court House in March, 1781. He was made a Brigadier General of Militia and served under Lafayette in the Battle of Jamestown, soon after which he died. Lafayette said of him "His services at King's Mountain and Guilford would do his memory everlasting honor and ensure him a high rank among the defenders of liberty in the American Cause", while Jefferson declared "Gen. Campbell's friends might quietly rest their heads on the pillow of his renown." His wife was a sister of Patrick Henry.

ARTHUR CAMPBELL

Arthur Campbell, a cousin of Gen. Wm. Campbell, was born in Augusta County, Va., in 1743. At fifteen he was captured by the Indians and carried to Lake Erie. Escaping, he was employed as a guide, receiving therefor one thousand acres of land which he located near the present site of Louisville, Ky. In 1772, he was a Justice of Fincastle County, Va., and later a Major of Militia. After the Battle of Point Pleasant, 1775, he represented his county in the Virginia Assembly. In 1776, he was chosen County Lieutenant for Washington County, which office he held for thirty years. He joined Sevier in the movement to establish the state of Franklin for which Patrick Henry removed him from office and the legislature re-instated him In his latter life he joined his sons at Middlesburg, Ky., where he died in 1811.

JOHN CAMPBELL

Captain John Campbell, a younger brother of Arthur, a lieutenant at Point Pleasant, was a captain at the Battle of Long Island Flats (1776) and served as County Clerk of Washington County from 1779 to 1815.

JOSEPH MAYSE

Joseph Mayse, who participated in the Battle of Point Pleasant, was from Bath County, Virginia. In April, 1840, he died, being in his 89th year. He had served as a magistrate in his district for nearly fifty years. He was a man of such remarkable memory he was considered an "official record." He was twice high sheriff of Bath County. His health was always perfect and he boasted he had never taken a dose of medicine.

GEN. ANDREW MOORE

Andrew Moore, a lawyer by profession, was born in Conniscello, Augusta, now Rockbridge, County, Va. In 1774, he was admitted to the Bar. In October of that year he was with Andrew Lewis's Army at Point Pleasant. In 1776, as a Lieutenant in the patriot army, he participated at Saratoga, where he was promoted to a captaincy and served there- after three years. The Virginia Legislature made him brigadier general of militia and in 1808 major general.

Gen. Moore had the distinction of being the only man west of the Blue Ridge, prior to the civil war, who ever represented Virginia in the Senate of the United States; which was during the administration of President Jefferson. He was a member of the Virginia Assembly from 1781 to 1789 and again from 1798 to 1800. In 1788, he was a delegate to the convention which ratified the United States Constitution. He was a member of Congress from Virginia during the entire administration of President Washington. In 1800, he was elected to the United States Senate, where he served three years. In 1810, President Jefferson appointed him United States Marshall for the state of Virginia, which office he was filling at the time of his death. His son, Samuel Moore, represented Virginia in the Legislature and in Congress, a member of the Virginia Constitutional Convention 1829. In 1861, Samuel Moore opposed the secession of Virginia, but, going with his state, he served in the Confederate Army. The family have always been distinguished.

GEORGE MATHEWS

George Mathews was born in Augusta County, Virginia, in 1739, and died August 30, 1812. At twenty-two years of age he led a volunteer company against the Indians. He was in command of a company of Augusta troops at Point Pleasant, Oct. 10, 1774, and participated with the patriot army throughout the Revolution. He was engaged at Brandywine. At Germantown he received nine bayonet wounds, was captured with his whole regiment and confined in a prisonship at New York until December, 1781. He then joined Gen. Nathaniel Greene's army in command of the Third Virginia Regiment. In 1785, he removed with his family to Georgia and settled in Oglethorpe County. In 1781- 1791 he represented Georgia in Congress; was Governor of Georgia, 1793-6. He was brigadier general of Georgia. In 1811, he was authorized by the President of the United States to take possession of West Florida and captured Amelia Island. His son, George, became a Supreme Judge of Louisiana. He admitted no superiors but Washington. He was short, stout, erect, features bluff, hair red, complexion-florid. He died when on a visit to Washington and is buried in St. Paul's churchyard. His four children, were Mrs. Andrew Barry, of Staunton, Va., Mrs. Gen. Samuel Blackburn, and Mrs. Isaac Telfair, of Staunton, and one son, Judge George Mathews, above mentioned. He was three times married, (1) to Miss Amelia Paul, (2) to Mrs. Margaret Reed, of Staunton, and (3) to Mrs. Flowers, of Mississippi. He was divorced from his second wife.

SAMPSON MATHEWS

Sampson Mathews, Commissary of Col. Charles Lewis's Regiment, was called "Master Drover of the Cattle." In 1756, Deputy Sheriff Sampson Mathews assumed the functions of Chancellor of Augusta County. In 1764, he was appointed a Justice of the Peace for Augusta. In 1776, with Alexander St. Clair, he was appointed by the state of Virginia as trustee "to erect at public expense and superintend a manufactory at such place as they may think proper for the manufacture of sail duck,' this preparatory for equipment of a Virginia fleet for Revolutionary service. He became Col. Sampson Mathews of the Revolution. In 1781, he commanded the regiment that repelled Arnold's invasion of Virginia. He was one of a committee to draft instructions for the members of the Virginia Convention at Richmond, Feb. 22, 1775.

COL. JOSEPH CROCKETT

Nothing can be truer than that God provides men for the hour. Among the one hundred men who participated in the Battle of Point Pleasant, and went on to become settlers of Kentucky, that state so open to the prey of the Indians that its first three years saw more people slaughtered by Indians than that state had white population at the end of that time, had among its other emigrants who were in the Battle of Point Pleasant, Col. Joseph Crockett, for a sketch of whose life we are indebted to his illustrious grandson, Col. Bennett H. Young, of Louisville, Ky. It is as follows:

"My great grandfather, Col. Joseph Crockett, was born in Albemarle, and was one of the men who marched with Andrew Lewis, with Charles Lewis, and with William Russell, and was engaged in the conflict at Point Pleasant. He was then a young man.

"He returned shortly after the battle of Point Pleasant, and remained for several months as First Lieutenant in a company of Colonial Militia that was stationed at Point Pleasant. When the fires of the Revolutionary War were kindled, and it was necessary for every man to go to the front to resists British invasions, the Indians were to be left a little while to themselves. Virginia and North Carolina were then sending their pioneers to make the settlements of Kentucky which were, under God's providence and direction, one of the chief agencies in the success of the colonies in their great struggle against the mother country. My ancestor organized a company and marched to the front. He was successively a Captain, Major and Lieutenant-Colonel. He was a Major in Morgan's riflemen, and recruited two companies for that celebrated organization. He was a Lieutenant at White Marsh, was a Captain when Burgoyne surrendered, was engaged in all the battles previous to that great event. He was at Princeton and Trenton, and Valley Forge and Red Bank, and in 1779, he raised a regiment known as the Crockett Regiment, which was sent west to assist George Rodgers Clark in his war with the Indians, and was with that distinguished soldier, second in command, in all the skirmishes and battles with the northwestern Indians on the Ohio and Miami Rivers, and helped to destroy Chillicothe, and the Indian towns on the Wabash, and throughout the northwest, and in these battles stayed the uplifted hand of the cruel and avenging Indian, who would otherwise have wreaked his cruelties upon the frontier settlements of both Pennsylvania and Virginia and thus coming in the rear of these enemies, avenged many of the wrongs heaped upon the Virginia, Pennsylvania and Kentucky settlements."

JAMES ROBERTSON

James Robertson, (by some authors written Robinson), with Val. Sevier, discovered the Indians before the Battle of Point Pleasant. He was born in Brunswick County, Va., 1742, died in Chickasaw County, Tenn., Sept. 1, 1814. He was the personal friend of Daniel Boone. He did more to consummate a peace between the Indians and whites than any man in Tennessee, when he became the founder of Nashville, where he withstood, with a handful of men, a siege of one thousand Indians. Flattering offers were made him by the Spanish government to cut the territory of Tennessee loose from the government, and, with Watauga and Kentucky, establish an independent country which he indignantly declined. In 1790, he was appointed Washington. He shared with Sevier the honors and affections of Tennessee.

JOHN SMITH

Ensign John Smith, of the battle of Point Pleasant, left sons, Abraham, of Rockingham Co., Joseph and Silas H. of Augusta Co., and daughter Nancy, who married Wm. Crawford, His family have been distinguished.

BENJAMIN HARRISON

Benjamin Harrison commanded a company at Point Pleasant In July, 1775, he was appointed Captain of a company of Minute Men. In 1778, he was appointed Lieutenant Colonel of Militia for Rockingham County. He was a native of Loudon County, Virginia. He was the founder of Harrisonburg, Va. He died in 1819.

HUGH AND JAMES ALLEN

Hugh Allen was a Lieutenant in Col. Charles Lewis's Regiment at Point Pleasant, where he lost his life and was buried beside Col. Lewis. His three sons, John, William and Hugh, all settled in Kentucky. His brother, James Allen, who lived eight miles from Staunton, was Captain of Militia in 1756 and was in the battle of Point Pleasant, and witnessed the death of his brother. He died in 1810, aged ninety-four years and was an elder in the Augusta Stone Church for sixty-four years.

JUDGE SAMUEL MCDOWELL

Judge Samuel McDowell who, as Captain McDowell, commanded a company of Augusta troops at the Battle of Point Pleasant, was a native of Rockbridge County, Va. He married Mary MeClung, and, with his seven sons and two daughters, in 1784, emigrated to Danville, Ky. He was one of the first Kentucky Court and was a member of the convention that framed for Kentucky the first Constitution. He presided over nine political conventions which convened in Danville from 1784 to 1790.

In 1776, he was a member of the Virginia House of Burgesses, representing Rockbridge County. As Col. McDowell, he commanded a Rockbridge Company during the Revolution. On April 20, 1781, he wrote the Governor of Virginia, when a draft had been ordered from his county for April 26th, that if the men were drawn the county would be ruined, as two thirds of the men had been engaged in the services all the time and there were no new ones to put in the crops, and that he had marched with 200 men to join Gen. Greene before the battle of Guilford Court House.

JOHN SEVIER

John Sevier, born in Rockingham Co., Va., Sep. 22, 1745, died near Fort Decatur, Ga., Sept. 24, 1815, was educated at Fredericksburg, Va., married at 17 years of age and became the founder of New Market on the Shenandoah. In 1772, he was appointed Captain of the Virginia line and moved to Watauga. In the Dunmore War, he resumed his rank in the "Virginia line and participated in the Battle of Point Pleasant. When what is now Tennessee was organized into Washington District, North Carolina, John Sevier was chosen a delegate to the legislature. In 1777, he again represented Watauga and procured for his state, courts and rights of extension. He was appointed clerk of the court and district judge and, with his friend, James Robertson, was in control of the judicial and administrative functions of the settlement. He was elected colonel and enlisted without exception every able bodied man between the ages of 16 and 50. With Col. Isaac Shelby he planned the battle of King's Mountain. He continued to command the forces against the Indians. When the new state, Franklin, afterwards Tennessee, was organized, he took the oath of Governor March 1, 1785. When the new state became a part of the Union, he was the first representative to Congress from the valley of the Mississippi, 1789-1790, and in 1796, when Tennessee became a state, he was elected its first Governor which office he filled for three years. He three times represented Tennessee as a state in Congress. He was in the active service of his country from the age of 17 years to 70. As long as he lived he was the real seat of power in Tennessee. A monument in Nashville attests to his memory and Sevier County in Tennessee commemorates his fame.

VALENTINE SEVIER

Valentine Sevier, who was a sergeant in Evan Shelby's company and a younger brother of Gen. John Sevier, in 1779, was a Captain in the Chickamauga Campaign and led a company against the British in North Carolina, 1780, which culminated in the Battle of King's Mountain. He rose to the rank of Militia Colonel and died at Clarksville, Tenn., in 1800.

JAMES HARROD

James Harrod who had been in Kentucky in the spring and summer of 1774, was with Col. Charles Lewis. He built the first cabin in Kentucky. He became Col. Harrod of the Revolution. He was a member of the first Kentucky Legislature. He was an able assistant of Geo. Rodgers Clark in securing ammunition. He declined to accept the appointment of Major of the first Regimental Militia of Kentucky. Harrodsburg Kentucky is named for him. A man named Bridges with whom he had had litigation murdered him, the date is thought to be July, 1793.

WILLIAM RUSSELL

William Russell was but fifteen years of age when participating in the Battle of Point Pleasant. During the Revolution he rose to the Rank of Colonel. He was a Lieutenant at the Battle of King's Mountain and Guilford Court House. After the Revolution he settled in Fayette County, Kentucky. He was appointed to command a regiment in the regular army. In 1792, he represented Fayette County in the Kentucky legislature, which was repeated a dozen times until 1825, when he contracted a cold at a public meeting where he was called to preside, which resulted in his death.

JAMES MONTGOMERY

Captain James Montgomery, who was in the Battle of Point Pleasant, settled in Kentucky, as did James Knox, who was in Isaac Kite's Company. Others of Isaac Kite's Company who settled in Kentucky were James McCullosh, John Shelp, William Field, Thomas Glenn, David Williams, James Brown, John Cowan, John Wilson, Abraham Chapline and John Clark.

JOHN CRAWFORD

John Crawford represented Montgomery County, Kentucky, in the Legislature in 1812.

WILLIAM CHRISTIAN

While Col. Wm. Christian was not an actual participant in the Battle of Point Pleasant, he, with his three hundred volunteers troops, not arriving until 11 p.m., the night of the battle, yet they did noble services to the bleeding army and the valiant dead. He was a native of Augusta County and educated at Staunton. He participated in the Braddock campaign. He married a sister of Patrick Henry and settled in Botetourt County. In 1775, he settled in Kentucky and Christian County is named in his honor. In April, 1776, he was killed by a party of Indians who had stolen his horses and in whom, with a party of friends, he was in pursuit.

GEORGE SLAUGHTER

Col. George Slaughter, a son-in-law of Col. Field, after the Battle, settled in Kentucky and was one of eight delegates to Congress out of the city of Louisville.

JAMES TRIMBLE

James Trimble, a participant in the Battle of Point Pleasant, aged then but eighteen years, had in 1770, been a prisoner of the Indians. In 1780 or 81, he emigrated to Woodford, Ky., being one of the earliest settlers. In 1804, he died in Kentucky, having made preparations to move to Hillsborough, Ohio, where his family removed after his death. The Trimbles became eminent. Allen Trimble became Governor of Ohio. Wm. A. Trimble was a Major in the War of 1812, and in 1819 a Lieutenant Colonel in the Regular Army and a United States Senator from Ohio.

JOHN DICKENSON

Captain John Dickenson, who commanded one of Col. Charles Lewis Companies of Augusta County troops and who was wounded during the Battle of Point Pleasant, was left with Col. Fleming at the fort when Andrew Lewis, with his troops, advanced into Ohio. In 1785, he surveyed 502 acres of land at the mouth of Campbell's Creek on the Kanawha River. In 1777, he commanded, as a Colonel with Major Samuel McDowell, the Botetourt troops, as well as troops, from Augusta County, and marched to the defense of the fort at Point Pleasant, accompanied by Capt. Hall's company of Rockbridge volunteers, numbering in all about 700 men, and he witnessed the murder of Cornstalk, which with the officers in command he tried to prevent. In 1791, when Bath County was organized, he was appointed one of the first gentlemen justices of that county which honor he declined. He died in 1799, owning large tracts of land in Bath, Greenbrier and Kanawha Counties, besides large holdings in Kentucky. He left sons, Adam and John and daughters Nancy, married Joseph Kinkcaid, Mary, married Samuel Shrewsbury and Jane, who was the wife of Charles Lewis son of Col. Charles Lewis, some of the descendants of the latter still being residents of this county.

ANTHONY BLEDSOE

Anthony Bledsoe, born in Culpepper County, in 1733, in 1774, moved to the neighborhood of the Shelby's. He was a magistrate of Botetourt, Fincastle and Washington Counties, and a member of the Virginia Assembly from Washington County, 1777-78. He moved to Bledsoe's Lick, North Carolina, and represented his district in the assembly of his state from 1785, to 1788, when he was killed by Indians. He was in charge of the commissary under Col. Christain at Point Pleasant with the rank of major. He commanded the forces at Long Island until July, 1777, and in 1779, went out against the Chickamaugas and did not participate in the Battle of King's Mountain because he felt it was his duty to remain at home and protect the frontier.

WILLIAM COCKE

Captain William Cocke, of the Battle of Point Pleasant, served in the legislature of four states and in the federal senate, as well as was prominent in his military career. He was born in 1748, in Amelia County, Virginia He studied law and removed to the Holston Valley. He was a brilliant orator and popular. After the Dunmore War he settled for a while in Boonesborough, Ky., but returned to the Watauga and participated in the Cherokee Campaign. In 1777, he was a member of the Virginia Assembly. In 1780 he led a company at King's Mountain. He was made a Brigadier General of Tennessee and, in 1796, one of the federal senators and was re- elected in 1799.

In 1809 he was elected a circuit judge of Tennessee. In 1812 he removed to Columbus, Miss., and enlisted as a private in the war of that year and soon rose to distinction. He died in 1828 at Columbus, Miss.

JOHN SAWYER

Col. John Sawyer, born in Virginia 1745, died in Knox County, Tenn., in 1831, and was with the Shelby's at Point Pleasant. In 1776, he served in the Cherokee Campaign and in 1779 in the Chickamauga expedition and commanded under Gen. Shelby a company at King's Mountain. He was a Major, next a Colonel and was a representative to the state assembly, of Tennessee.

JOSEPH HUGHEY

Joseph Hughey, of Shelby's Company, was killed when attempting to bring the news of the Indians' presence to camp before the battle. James Mooney, who accompanied him, succeeded in reaching the camp, but was killed during the battle.

PHILIP LOVE

Capt. Philip Love later served as a Colonel in Christian's Cherokee Campaign in 1776.

ELLIS HUGHES

Ellis Hughes, who is thought to have been the last survivor of the patriots of the Battle of Point Pleasant, went, after Wayne's treaty, to Ohio and died March, 1845, at Utica, N. Y., where he was highly respected. He was buried with military honors.

Reared in his native state Virginia, nurtured amid the scenes of forest savagery, wherein by Indian depredations he lost his father and sweetheart, it is not surprising that he pursued the dusky foe until he had disappeared from Virginia and from his adopted home, Ohio.

JOHN STEELE

John Steele, who was wounded during the engagement at Point Pleasant, was born in Augusta County, Virginia, about 1755. He was an officer in the Battle of Point Pleasant and served throughout the Revolution. He was again wounded at the Battle of Germantown. He was for many years a member of the Executive Council of Virginia and in the administration of President John Adams, was a commissioner to treat with the Cherokee Indians. From 1798 to 1801 he was Secretary of the Mississippi Territory.

AZARIAH DAVIS

Azariah Davis of the Battle of Point Pleasant, was a member of Harrod's Company. He was one of the members of the First Kentucky Legislature and is mentioned (1775) among the first settles of Harrodsburg, Kentucky

JOHN TODD

John Todd became one of the founders of Louisville, Ky. He was with Col. George Slaughter.

CHAS. E. CAMERON

Chas. E. Cameron and his brothers, Hugh and George Cameron, were with the Virginia troops at Point Pleasant in which engagement George Cameron was killed. They were brothers-in-law of Col. Charles Lewis, who was killed in that battle, whose wife Sarah Murry, was their half-sister. George Cameron resembled in person and being his distinguished father, Dr. John Cameron, of Staunton, who had emigrated from Scotland. Charles Cameron served throughout the Revolution, as a Lieutenant and was with the Virginia troops at the surrender of Yorktown In 1790 he was one of the gentlemen justices of Augusta County. On December 14, 1790, he received a land grant in Bath County, Va., where he located, about four miles from Warm Springs. He accumulated large land interests. His residence of stone was magnificent for its time and overlooked the Jackson River. Major Cameron became the first Clerk of Bath County, serving both courts for a number of years. After the Revolution, he became Colonel of Militia. As a personal friend of Gen. Lafayette, he was presented by him with a beautiful cane which he used and prized until his death, which occurred June 14, 1829. He was survived by his widow, Mrs. Rachel Primrose Warwick and one son, Andrew Warwick Cameron.

SILAS HARLAN

Silas Harlan, of Berkeley County, Virginia, was in Captain Harrod's Company and, after the Battle of Point Pleasant, emigrated to Kentucky. In 1779, he commanded a company of spies under Gen. Geo. Rodgers Clark in the Illinois campaign. Gen. Clark pronounced him one of the bravest and most accomplished soldiers who ever fought by his side. He was a Major at the Battle of Blue Licks, where he fell. He was but thirty years of age and unmarried.

JACOB WARWICK

Jacob Warwick, of Bath County, Virginia, on the morning of the Battle of Point Pleasant, had gone out early to kill beeves and prepare rations for the army. He and the men who accompanied him hearing the first shots of battle, thought Dunmore had arrived and that the guns were a salute. Later they thought it a practice exercise, but, determining to see for themselves, they joined the army in time to help materially in turning the tide of victory.

Jacob Warwick is buried beside his wife at Clover Lick Cemetery in Barth County, Va., where he died Jan., 1826, in his 83rd year. He died at the home of his daughter, Mrs. Major Charles Cameron, on Jackson River.

The VAN BIBBERS

The brothers, John, Isaac and Peter Van Bibber, and Jesse, son of Peter, were participants in the Battle of Point Pleasant. Mrs. Miriam Donley, a Van Bibber descendant, writing for the July, 1903, *West Virginia Historical Magazine*, says "Isaac had come from the Carolinas on a visit to his brother in Botetourt County, when the call to arms resounded through the land. Although a Baptist minister, he could and would not resist, as hearts were that day attuned to martial music, and he responded to its call. He fell mortally wounded besides Colonel Charles Lewis. Peter fought with such bravery he was promoted and complimented on the battle field. John Van Bibber was written of by all historians as Captain and family notes say he was made Captain after the Battle of Point Pleasant and Commissary of Fort Randolph. The Van Bibbers continued to defend the border although Isaac, the son of Isaac, fell at Point Pleasant, while Jacob and Mathias Van Bibber died twenty years later. As late as 1843, Captain Jesse Van Bibber was still residing on Thirteen Mile Creek in Mason County, now West Virginia. He with his brother, John Van Bibber were among the earliest settlers of that County."

Howe, the Historian, who in writing the *History of Virginia* in 1836, said "There is living upon Thirteen Mile Creek, Mr. Jesse Van Bibber, and aged pioneer in this county. His life, like his own mountain stream therein, was rough and turbulent at its commencement; but as it nears its close, calm and peaceful, beautifully reflecting the Christian virtues."

LEONARD COOPER

Captain Leonard Cooper, another Revolutionary soldier who is buried in Mason County, West Virginia, participated in the Battle of Point Pleasant. Prior to the Revolution, he held a commission in the Colonial army of Maryland. Learning of Dunmore's War, he hastened to Staunton, Virginia, and entered the Army of General Lewis. He remained in the service until the close of the Revolutionary struggle. In 1789 Major Cooper removed with his family from Maryland to Fort Randolph, later erecting a Block House, known as Cooper's Block House, (where Mr. George W. Pullin now resides) in Cooper District, nine miles from the mouth of the Kanawha, on the upper side. He there removed with his family.

In 1804, when the new County of Mason was organized, Major Cooper was appointed a justice of the peace in which capacity he served until his death which occurred in 1808. His remains were buried near his home. His son, Leonard, born in 1791, was the first white child born at Point Pleasant. Another of Leonard Cooper's children, Mary became the wife of William Trotter, son of Richard Trotter, killed in the Battle of Point Pleasant and Anne (Trotter) Bailey, who, going from Cooper's Block House, by canoe to Gallipolis, where in 1800, their marriage ceremony was performed. This is said to have been the first Virginia marriage performed in the French Settlement.

WILLIAM ARBUCKLE

Captain William Arbuckle, of Rockingham County, Virginia, deserves to rank with Daniel Boone and Simon Kenton in the valor displayed in wresting from savagery the vast domain in which his expedition laid He was not only with General Andrew Lewis at the Battle of Point Pleasant, where as a pilot (having first visited the mouth of the Kanawha in 1764) he safely conducted that wing of the army, but when Geo. Rodgers Clark was organizing his expedition against the French Forts in Illinois from which the Indians were known to receive supplies, he (Capt. Arbuckle) tendered his services which were accepted and he acquitted himself with credit in that ever memorable campaign. He defended the fort at Point Pleasant. He married Catherine Madison, widow of Capt. Robert McClannahan, who fell in the Battle of Point Pleasant. He remained in command of Fort Randolph until 1795 when Wayne made his treaty with the Indians, when he bought land and located on the Kanawha four miles below the present town of Buffalo, where he and his wife passed a peaceful and honored old age. Among their descendants yet on the Kanawha are the families of Arbuckle, Craig, Alexander, Miller and others. William Arbuckle had two children born within the fort at Point Pleasant. He and his wife both are buried in the church yard at the Arbuckle Church in Mason County West Virginia, Simple stones are thus engraved:

"Wm. Arbuckle, born March,

1752, Died March 21, 1836,

Aged 84 years."

"Kitty Arbuckle, Died July 18,

1818, Aged 64 years."

JOHN YOUNG

John Young- became a settler in the present Kanawha County, and, in the military organization of the County, was a lieutenant of militia. He left a son, Jos. Young, from whom descend many residents of the valley.

JOHN HENDERSON

John Henderson, about 1740, with his brothers James and Samuel, came to Augusta County, Virginia, from Scotland. Descending from James, John his second son, was born 1740, and died March 24, 1787. In 1765 he married Ann Givens, sister to the wife of Gen. Andrew Lewis, and buying 300 acres of land, he settled in Greenbrier. In 1786, he was granted by Governor Randolph 350 acres, now in Greenbrier County, and 1400 acres on the South Side of the Kanawha in what is now Clendennin and Arbuckle Districts, Mason County, West Virginia.

In Greenbrier County, he became a Lieutenant of Militia and ranked as such in Captain Herbert's Company at Point Pleasant. Later he was Captain of the Greenbrier Militia and later was a Corporal in Capt. Gregory's Company in Daniel Morgan's Virginia Regiment, serving until April, 1779. In 1780, he was elected a Justice of the Peace, of Greenbrier County, which office he held in 1787, the time of his death. He was survived by his widow who died May 28, 1819, and children, Samuel, John, Margaret, James, Jean and William. John and Samuel inherited the lands on Kanawha, where they made permanent homes. Samuel building his cabin home at the mouth of Kanawha and in 1810 burned the brick and erected a commodious brick house, the second one in the county, now occupied by his granddaughter, Mrs. Ella M. Henderson Hutchinson and family.

John Henderson second, son of Capt. John Henderson, was a man prominent in the public affairs of Mason County, and he occupied and inherited that part of the tract of land adjoining his brother Samuel but running further up the Kanawha.

LUMAN GIBBS

Luman Gibbs was but 16 years of age when, with the army of General Andrew Lewis, he participated at the Battle of Point Pleasant. He was left as a part of the Garrison at the Fort. He became a noted scout and for twenty years he served in that capacity, wandering over the hills of the present County of Mason. His weekly route proceeded from Fort Randolph up the Kanawha to the Mouth of Eighteen Mile Creek, thence across to Letart Falls, thence down the Ohio to Point Pleasant, and his "All's Well" for twenty years dispelled the fears of the early settlers in and about the tort. The early settlers knew the route as "Gibb's Track." He married and located permanently in Mason County, where he has many descendants.

He had emigrated to Augusta County Virginia in 1755, coming from New Hampshire where he was born. He engaged at once in the Colonial Army in that year with Andrew Lewis in the Braddock campaign and again enlisted in his command for the Point Pleasant Campaign. He was as noted for his sunny disposition as for his bravery. He lived to a great old age and died 1837 and is buried in the Gibbs family burying ground eight miles from Point Pleasant.

In the same grave yard are buried Revolutionary soldiers James Ball and Isaac Robinson who too participated in the Battle of Point Pleasant and became residents of Mason County.

GEORGE EASTHAM

George Eastham, of Fauquier County, Va., who was in one of the companies with Col. Field at the Battle of Point Pleasant, was born in 1758, and hence was but a youth when engaged in that battle. He participated in many struggles throughout the Revolution. He married (1) Susan Woodside, who bore him nine children, among whom was Col. Lawson Eastham; his second wife, Mrs. Mary Brown, widow of James Brown, bore him three children, viz., Lucinda, Albert G. and Saunders. In 1817, he moved to Arbuckle district, Mason County, Virginia, known as

Five Mile Creek, and in the following year died. His son, Albert G. Eastham, born in 1805, father of a large family in Mason County, died Feb. 23, 1890, at his home in Arbuckle District being the last real son of "The Revolution" in the county of Mason. He left many descendants in that County who do honor to his name.

JOHN STUART

Col. John Stuart was the son of Col. David Stuart, County Lieutenant of Augusta County, when that county extended from the Blue Ridge to the Mississippi, 1755.

John Stuart, son of Daniel and Margaret Stuart, was born in 1749, in Augusta County and emigrated to Greenbrier in 1769 and built a house of hewn logs two and a half stories high, which he used as a residence and fort, known as Fort Union.

When his cousin Andrew Lewis rested his army at Fort Union and was ready to continue the march to Point Pleasant, his forces were augmented by Col. Stuart's and one company commanded by Capt. Robert McClannahan.

At Point Pleasant Captain Stuart's Company was one of the three sent up Crooked Creek in the flank movement that successfully put Cornstalk to rout.

Col. Stuart did not go on with the further battles of the Revolution, but continued the defense of Fort Union and organized a force and went to the successful relief of Fort Donnally when that fort was so vigorously attacked by the Indians.

November 25, 1780, Col. John Stuart became clerk of Greenbrier. At the close of the first deed book he makes valuable historical notes including an account of the Battle of Point Pleasant.

Col. Stuart married Agatha, the widow of John Frogg, killed in the Battle of Point Pleasant, she the daughter of Thomas Lewis, hence already his kinswoman. They had four children, Margaret, who married General

Andrew Lewis, son of Col. Charles Lewis; Jane married Robert Crockett; Charles A., married Elizabeth Robinson, and Lewis, who married, Sarah, the second daughter of Col. Charles Lewis.

Col. Stuart became one of the best business men and largest land owners of Greenbrier County. In the splendid stone mansion he had built, he lavishly entertained. Here were wont to meet the most intelligent, polished and distinguished men, not only of Virginia, but of other states and nations, and his generosity was only bounded by the demands of his neighborhood,

[In] 1788, he was a member of the Virginia Constitutional Convention. In 1793, he was appointed Lieutenant Colonel of the 79th Regiment of Militia. In 1776, he and his wife each contributed 500 pounds sterling to build the old stone church at Lewisburg, yet beautifully preserved. He was a member of seven literary societies including the American Philosophical Society. His library was extensive and valuable. He built in his own yard the first clerk's office of the county which is still standing. He presented the county the lot upon which the first court house at Lewisburg was built. He died August 23, 1823, and is interred in the old family burying ground.

THOMAS POSEY

Thomas Posey was born on the Potomac River in Virginia July 9, 1750. He early participated with the Virginia militia and with the rank of Captain, was Quartermaster to the Army of General Lewis.

In 1775, he was appointed a member of the Committee of Safety and that year raised a company which he commanded and assisted Gen' Andrew Lewis in driving Governor Dunmore from Gwinn's Island, July 8, 1776. In 1777, he joined the Continental Army at Middle Brook, N.J. Here he became one of the distinguished picket men of Morgan's Riflemen, distinguishing himself at Piscatoway, Bennington Heights and Stillwater. In 1778, he was promoted to the rank of Major, in command of the Morgan Riflemen. In 1778, he commanded the 11th Virginia Regiment, At the close of that year he entered the artillery service and was in charge of a battery under Wayne in the attack upon Stony Point, one of the most thrilling incidents of the Revolution, being the first field officer to enter the enemy's works. He witnessed the surrender at Yorktown. He retired with the rank of Brigadier General, settling at Spotsylvania Court House,

Virginia. [In] 1793, he removed to Kentucky, where he was elected Lieutenant Governor and, as such, President of the Senate. Moving to Louisiana in October, 1812, he was chosen to fill a vacancy in the United States Senate, President Harrison appointed him Governor of Indiana Territory which honor he declined. He was agent of Indian affairs from 1813 to 1816. He died at Shawnee Town, Ill., March 19, 1818. His first wife was the daughter of Colonel Sampson Mathews, of Virginia; his second wife, widow of Major Geo. Thornton, and daughter of John Alexander.

Posey County, Indiana, commemorates his name which name adds lustre to the roll of the army of General Lewis.

JOHN LEWIS

Major John Lewis, a nephew of General Andrew Lewis with whom he was engaged at Point Pleasant, died in 1823, at his home at Sweet Springs. He was the son of Wm. Lewis, brother of Gen. Andrew Lewis. He was noted for his courage, integrity and high sense of honor and continued in the service of the Colonies throughout the Revolution. As a Lieutenant he was engaged at Monmouth, Saratoga, Trenton and spent the winter of 1777 at Valley Forge. He rose to the rank of Major, which rank he held at Monmouth. In 1783 he returned to his Virginia home, but was much engaged on the frontier until the close of Wayne's Campaign. He was five feet, ten inches high, compactly built, muscular, strong and courageous. At the time of his death, he was an Elder in the Presbyterian Church.

WILLIAM CLENDENNIN

William Clendennin was a private in the Battle of Point Pleasant. Later he was commissioned Major in the Kanawha Militia. He represented Kanawha County in the Virginia Assembly in 1796-1801-1803. He was a Collector of Levies 1792-1793-1794. He was a justice of the peace and member of the first Court in the County, held at his house in 1789. In 1790, he settled in what is now Mason County, W. Va., In 1804, he carried the petition to the Virginia Assembly asking for the organization of Mason County Virginia and in 1805, was the first representative of that county.

Major Clendennin had settled about 1797 at Eight Mile, Mason County. In 1802 he purchased a part of the Hugh Mercer tract and built the first log cabin in Clendennin District, Mason County, and many of his descendants are living in Mason and adjoining counties in West Virginia and Ohio. By his son Charles, whose son William married Sophia Neale of Gallipolis, their son, James B. Clendennin, is survived by a daughter, Mrs. George Wallis, of Apple Grove.

Sophia, daughter of William Clendennin, married John Miller, and her sister Ann, married Henry Miller, both of Gallipolis. Another sister, Elizabeth, married John Bing of Gallia County, Ohio, from whom descended a large family. Their second child, Martha Young Bing, born in Gailia County, Ohio, Oct. 24, 1805. died Oct. 30, 1900, was the ancestor of the Filson and Cable families of Mason County, West Virginia.

ARCHIBALD CLENDENNIN

Archibald Clendennin, brother of William and George, married Nancy Ewing and lived on a farm a mile from Lewisburg. The family were attacked by Indians, and Archibald Clendennin was killed. His wife was captured by the Indians, but made her escape.

BENJAMIN LOGAN

Benjamin Logan was born in Augusta County, Virginia, in 1752. He emigrated, to Pennsylvania from Ireland, when a child, and when but fourteen, emigrated with his parents from Pennsylvania to Virginia, where his father died. By the law of entail then prevailing in Virginia he became the heir of his father's estate, but he divided it with his mother, brothers and sisters. He married and settled on the Holsten river and was with Col. Henry Boquet in his expedition against the Indians. He was in the battle of Point Pleasant. In 1775 he emigrated to Kentucky with Daniel Boone and established Logans Fort, where he moved with his family the following year. He was one of the most daring of Kentucky pioneers and his defense and relief of his fort is one of the most thrilling pages in Kentucky history. His expedition against the Indians at Chillicothe in which the Indians were put to rout and their supplies captured, including 150 horses, was admirably planned and executed. In 1788 he led a regiment of 600 men against the Indians of the North West. He passed his declining years in Shelby County, Kentucky, on his extensive farm, dying, Dec., 11, 1802. He was six feet three tail, powerfully built with nerves and courage like a lion. His son Wm. was the first white child born in Kentucky and became an eminent lawyer, being twice appointed appellate Judge of Kentucky and in 1820 was a United States Senator from Kentucky.

JOHN LOGAN

John Logan brother of Benjamin was engaged in the Battle of Point Pleasant. He emigrated from Virginia to Kentucky where he was a military leader and several times was a representative.

GEORGE CLENDENNIN

George Clendennin who participated in the Battle of Point Pleasant, represented Greenbrier County in the Convention at Richmond, 1788, that ratified the Federal Constitution of Virginia. In that year, he purchased 1030 acres of land, the site of the present city of Charleston, and in that year built Fort Lee, afterward called Fort Clendennin. In 1794, the town of Charleston was laid off, which was not named, but finally called Charlestown, in honor of Charles Clendennin, father of George.

The first court was held in Kanawha County, Oct. 5, 1789, at the residence of George Clendennin, a County Lieutenant. He was one of the first representatives of Kanawha County, 1790-1791- 1792 1794-1795. In 1794 he was made a trustee for the newly laid off town of Point Pleasant.

His wife was Jemima, claimed by some to be the sister of Thos. Pawing, of Ohio, but which has been found to be an error. He died after 1795, when his name last appears as signing a deed and in 1797 his wife appeared in court as his widow, when she gave bond as administratrix.

Parthena, daughter of George and Jemima, Clendennin married John Meigs of Marietta, Ohio. John Meigs dying, his widow married Major Andrew Bryan, their daughter Mary married John McCulloch, from whom descended Mrs. M. M. Moore, Mrs. P. S, Lewis, Mrs. J. J. Bright, John A. and Charles E. McCulloch, who were reared on a farm below Kanawha in Arbuckle District, Mason Co., West Virginia.

Mary, the third daughter of George Clendennin, married Major John Cantrell whose only daughter became the wife of the late C. C. Miller, of Mason county, who has left many descendants.

ALEXANDER BRECKENRIDGE

Alexander Breckenridge named for his maternal grandfather Alexander Breckinridge was in the Battle of Point Pleasant, and later served as Colonel in the 7th, Virginia in the Revolution, resigning in 1778. He was for many years Clerk of Augusta County. He and Patrick Henry married sisters.

CAPT. JOHN LEWIS

Captain John Lewis eldest son of Thos. Lewis was a nephew of Gen'l Andrew Lewis. He was born in 1749. He was wounded at the battle of Point Pleasant. He engaged in the struggles of the Revolution, was at Valley Forge and Jersey, and witnessed the surrender of Cornwallis.

STEPHEN TRIGG

Capt. Stephen Trigg, of the Battle of Point Pleasant, was a member of the Virginia Assembly from Fincastle in 1774, when Governor Dunmore dissolved that body. He signed the Articles of Association of the Colonies in 1775 and was active in protecting the frontier during the Revolution. In 1779 he emigrated to Kentucky and represented that county in the Virginia Assembly 1780, While leading a charge at the Battle of Blue Licks, 1782, he was killed.

WILLIAM HERBERT

Captain William Herbert was a Captain of Fincastle Militia, who participated in the Battle of Point Pleasant. He died 1776.

WALTER CROCKETT

Captain Walter Crockett was born on the South Fork of the Holston River. He was a county magistrate. He continued in the patriot army after the Battle of Point Pleasant and distinguished himself at King's Mountain, 1780.

JOHN FLOYD

John Floyd, who was a school teacher, made his home with Col. Wm. Preston, of Fincastle County, was a native of Virginia, born 1750. In 1774, he was appointed a deputy sheriff. In the spring of 1774, he led a surveying party to Kentucky and, returning, joined Wm. Christian in the

Point Pleasant expedition, arriving too late to actively engage in the battle, but was active in the good offices of his company in ministering to the needs of the army. He married Jane Buckhannon, niece of Col. Preston, and in 1779 located in Kentucky, where, in 1783, he was killed by Indians. His son, John Floyd, who was born in Jefferson County, Virginia, 1770, represented Virginia in Congress 1817-1829, Governor of Virginia 1829-1834. South Carolina cast her electoral vote, for him for president in 1832. His son, John B. Floyd, grandson of the John Floyd of Point Pleasant campaign, was a member of the Virginia Legislature 1847 and was Governor of Virginia 1850-1853 and was Secretary of War under President James Buckhannan. He was indicted by the government, charged with the misuse of government supplies and funds. He demanded a trial and was exonerated. He resigned his position and became Gen. John Floyd of the Confederate Army. He married Sallie Buckhannan, granddaughter of Wm. Campbell, of the Battle of Point Pleasant, and a niece of Patrick Henry. They had no children.

BENJAMIN LEWIS

Quoting from the biography of his descendant, State Historian and Archivist Virgil Anson Lewis, in *Men of West Virginia* (1903) page 31, "'His paternal ancestors were among the first settlers of the Shenandoah Valley, where they were founders of the city of Staunton. They were active frontiersmen and participants in the Revolutionary and Indian Wars. His great grandfather, Benjamin Lewis, was wounded in the Battle of Point Pleasant and after the wars were over in 1792 settled in what is now Mason County, West Virginia, and is buried in Waggoner District, near the spot where he thus found a home."

The following from the War Department Adjutant General's Office, Washington, D. C., under date March 28, 1908, is authoritative that after the Battle of Point Pleasant, he continued to serve in the patriot army: "It is shown by the records that one Benjamin Lewis who served as a sergeant in Capt. John Spotswood's Co. 10th Virginia Regiment Commanded successively by Col. Edmond Stevens and Major Samuel Harnes and Col. John Green, Revolutionary War. He enlisted November 29, 1776, to serve three years and was discharged July 5, 1778.

Signed, F. C, Ainsworth

The Adjutant General."

That Benjamin Lewis above referred to was not a descendant of John Lewis, the founder of Staunton, we quote from a memorandum of Mrs. Sarah Lewis Rodgers, who was raised at the old Lewis home on Muddy Creek, in Greenbrier County, who moved in pioneer days to Illinois. Writing to her nephew, Rev. Jacob H. Lewis, a Presbyterian minister, of Greenbrier county, the latter dying- at 92 years of age, the manuscript is yet preserved and says:

"Our Lewis family in Greenbrier county originated from three brothers, John George and Benjamin Lewis, who came to the county in an early day from the Valley of Virginia. About the close of the Revolution, Benjamin went to the Ohio. George Lewis never married. John married Miss McCrary and their sister, a Mrs. Van Orzel, is buried in the old Caraway graveyard."

None of the sons of John Lewis founder of Staunton Virginia left descendants such as those above described, but it has been claimed that the above Benjamin was the son of Thomas, he the son of John. Mrs. M. L. Price, West Virginia historian of the John Lewis family from whom she descends, says Thomas Lewis' son, Wm. Benjamin, was born 1778 (four years after the Battle of Point Pleasant, in which the family traditions and papers have always shown Benjamin Lewis to have been) while L. L. Lewis, of Richmond, recognized as an authority on the John Lewis, Staunton, branch, says: "Thomas Lewis had a son Benjamin, but he lived and died in Rockingham County, Virginia."

As early as 1812 we find in Mason county that Benjamin Lewis conveyed land which was acquired before Mason County was formed, as the land books show no transfer to him in that county prior to 1812 and he continued to buy and convey lands as the records show, and that in 1831 there was much conveying of titles of his lands by his children which would indicate that he died on or before that year. Land conveyances show the given name of the wife of Benjamin Lewis to have been Nancy, and their children to have been Sarah, who married Samuel Edwards, John who married --- Edwards, Benjamin, Jr., who emigrated to Iowa, Catherine who married Michael Newhouse, George who married Margaret Winkleblack, William who married Lucinda Clendennin, Andrew, Isaac and probably others.

JOSIAH RAMSEY

After being engaged in the Battle of Point Pleasant, Josiah Ramsey returned to Augusta County. He served as a scout in the Cherokee campaign of 1776. In 1778, he removed to Kentucky. In 1779, he moved to Cumberland Settlement, where he was appointed Major of Militia and was frequently engaged against the Indians. He lived to an advanced age spending the close of his life with a son in Missouri.

WILLIAM BOWEN

William Bowen, often related a hand to hand encounter with an Indian antagonist at Point Pleasant whom he finally overpowered. He was a native of Maryland, born 1744. In 1759 he engaged in the border warfare with Wm. Christian. In early life, he had moved to Augusta County, Virginia. In 1784 he removed to Summers County, Tennessee, where he passed the remainder of his life.

JOSEPH DRAKE

Joseph Drake who was with Wm. Christian's Regiment at Point Pleasant, had served as a private in Boquet's expedition in 1764. He was one of the Long Hunters 1770-71. In 1773 he married Margaret, daughter of Col. John Buchhanan. In 1775, he visited Kentucky and in June of that year led an exploring party on Green River. He resided at Abington, Virginia until 1778, when he moved to near Boonesborough, Ky., and in August of that year was killed by the Indians. He was a typical frontiersman.

WILLIAM EDMISTON

Lieutenant William Edmiston (Edmondson) a native of Maryland, born 1734, moved at an early date to Augusta County, Virginia. He was a private in the French and Indian War and the Cherokee Campaign 1760. In 1763 he was appointed Lieutenant of Militia for Augusta County. He was in Capt. Wm. Campbell's Company at Point Pleasant and was his second in command at King's Mountain, in which eight members of his family were engaged, three of whom were killed. One of those who survived of that family was James Edmiston who has descendants living in many counties in West Virginia, including the county of Mason.

WILLIAM INGLES

Major William Ingles, who, at the Battle of Point Pleasant, was in charge of a commissary with the rank of Major, was a native of Ireland, born 1729, emigrating with his father when a child to Pennsylvania, settling with John Draper at Draper's Meadows in 1748. In 1750 he married Mary Draper, whose capture by, and escape from, the Indians, is one of the thrilling pages of pioneer history. During the Indian Wars, Wm. Ingles was active in defense of the frontiers. In 1756 he was a Lieutenant in the Sandy Creek expedition. In 1758-60, he defended the fort at Ingle's Ferry. In 1777 he was made Colonel of Militia in the organization of Montgomery County. In 1782, he died at his home at Ingles Ferry.

THOMAS INGLES

Thomas Ingles was with his mother, Mary Ingles, who was captured by the Indians, remaining with them until 1768, practically becoming a young Indian in his habits. Returning to his home for a few years, he tried to adopt the habits of civilization and education, but he never forgot his Indian friends. He was in the battle of Point Pleasant, remaining the following winter in the Fort, during which time he visited the Indians at Scioto. In 1782, his wife was captured by the Indians and his home burned. He removed afterward to Tennessee, thence to Mississippi, where he died.

HENRY PAULING

Capt. Henry Pauling who commanded a company of Botetourt troops at Point Pleasant, continued in frontier service and in 1777 went with Col, Bowman to the relief of the Kentucky frontier soon after which he settled in that state and represented Lincoln County, Ky, in the convention of that state that ratified the Constitution of the United States, but he voted against the ratification of that instrument.

FRANCIS SLAUGHTER

Col. Francis Slaughter who was at the battle of Point Pleasant was a member of one of the best Virginia Families. He married a daughter of Robert Coleman of Dunmore and in 1785 moved to Kentucky, settling in Hardin County.

LAWRENCE AND GEORGE SLAUGHTER

Lawrence and George Slaughter each married a daughter of Col. John Field and both were in his regiment at the Battle of Point Pleasant. Col. George Slaughter in 1776 raising a company, participated in the 8th Va. Reg. at Brandywine and Germantown. In 1779 he joined Shelby in the Chickamauga Campaign and in that winter started to reinforce George Rodgers Clark, but was obliged to winter at Louisville, Ky., joining him the following June, after which he returned to Virginia and in 1784 represented his county in the house of Delegates. Later he moved to Jefferson county, Kentucky, thence to Charleston, Ind., where he died June 17, 1819.

THE MCAFEE BROTHERS

McAfee station on Salt River, in Mercer County, Kentucky, commemorates the name of five McAfee brothers, James, Robert, George, William and Samuel, who lived on Sinking Creek, Botetourt County, Virginia, from which place they finally all emigrated to Kentucky, in 1779. The first three named with James McCorn, Jr., and Samuel Adams, were Kentucky explorers in 1773 with Col. Bullit and Hancock Taylor. They all participated in the Battle of Point Pleasant.

William McAfee was in the George Rodgers Clarke expedition and was killed in 1780. George died in 1803 at his home on Salt River. Samuel died in 1801, James in 1814, and Robert who was one of the early dealers, who, by flat boat, took large cargoes of produce to New Orleans, in 1795, when on such a mission, was killed by a Spaniard in that city, who was attempting to rob him. They left descendants, many of whom are yet residents of Kentucky.

JAMES KNOX

Major James Knox served under Col. Chester in the capacity of scout in 1774. During the Revolution he commanded a company of Morgan's Riflemen engaged at Saratoga and Stillwater, returning with the rank of Major. Settling in Kentucky, he married Mrs. Logan, the widow of Benjamin Logan, who was in the Battle of Point Pleasant. James Knox died in 1822. He had accumulated a good fortune and was respected by all who knew him.

JOHN MADISON

John Madison was of the distinguished Virginia family that gave to America the president of that name, being a first cousin of President Madison. His son James Madison, was the first American Episcopal Bishop. Other of his sons who distinguished themselves were Thomas, Rowland and George, who emigrated to Kentucky. John Madison was the first Clerk of Augusta County and represented that county in the Virginia Assembly in 1751-52. He married a Miss Strother, sister to the wives of Thomas Lewis and Gabriel Jones of Augusta County

ELIJAH KIMBERLING

Elijah Kimberling of Bath County, Va., who was engaged in the Battle of Point Pleasant, returned to Bath County, Va., where he resided until the time of his death. So pleased, however, were his sons with his description of the Kanawah Valley, that his four sons, Joseph, James, Jacob and Nathaniel emigrated to Mason County West Virginia and located on farms in Union District, near Arbuckle Postoffice. They became the progenitors of a large and influential family in the Kanawha Valley. Among whom were Elijah Kimberling, for many years a public official of Mason County, who married Margaret Catherine Jones, a native of Culpepper County, Virginia, daughter of Joseph Jones, and Ann Winn, his wife.

WILLIAM EWING

William Ewing, a member of Arbuckle's company at Point Pleasant, settled on Swago Creek, tributary to the Greenbrier, near Buckeye, Va. He was one of the garrison at Point Pleasant and witnessed the murder of Cornstalk.

WILLIAM MCKEE

William McKee, born in Ireland in 1732, and emigrating when a youth to the Valley of Virginia, was in the Braddock Campaign. At Point Pleasant he was a lieutenant in Captain Murray's company. He later represented Rockbridge County in the Virginia Legislature and voted in favor of the adoption of the Constitution. He emigrated to Lincoln County, Kentucky, where he died in 1816.

CHARLES SIMMS

Charles Simms was in the division commanded by Col. Lewis who expired in his arms. In the continental army he was first major of the 12th Virginia, later Lieut. Col. of the 6th Va. and later of the 2nd Virginia Regiment. On Dec. 7, 1777, he resigned from the army and practiced law at Alexandria, Virginia, where he continued to reside until the time of his death.

GOERGE MOFFATT

Captain George Moffatt was born in 1735. His father was killed by Indians in 1749, enroute to South Carolina. In 1763, George was Captain of a company of Rangers in pursuit of Indians that had killed his step-father, John Trimble, and captured his sister and half-brother. He rescued his loved ones. Col Wm. Christian was his uncle and Samuel McDowell his brother-in-law and in the battle at Point Pleasant, with him were many of his kindred. In the war of the Revolution he was active and commanded a regiment at Guilford Court House. From 1781-83 he was County Lieutenant of Augusta. He died at his home eight miles northwest of Staunton in 1811.

JOHN MURRAY

Capt. John Murray, killed in the battle of Point Pleasant, was a brother in law of Col. Charles Lewis, and a half-brother of Charles Cameron, and Geo. Cameron, the last named, was killed in the battle.

WILLIAM TROTTER

William Trotter who was engaged in the Battle of Point Pleasant, was an Englishman by birth, coming to America and an indentured servant. Coming at the same time with Ann Hennis, who, like him, was "sold out" to defray the expense of their passage. They were bought in Augusta County, Va., and when his term of service had expired, he enlisted with Col. Charles Lewis, and was killed in the Battle of Point Pleasant. This so incensed his widow that she donned a semi-male attire and with rifle and tomahawk she was seen at musters like a man. Later she married James Bailey.

JAMES BAILEY

James Bailey, who was with Col. Charles Lewis and who afterward married Ann Hennis, the widow of Richard Trotter, is more noted for being the husband of Ann Bailey, the heroine of the Kanawha Valley than for his own achievements. He was assigned to garrison duty at Fort Clendennin where the present city of Charleston is now located. Ann Bailey was the mother of one son, William Trotter, who located in Harrison Township, Gallia, Ohio, where his descendants yet reside.

WALTER NEWMAN

Walter Newman, a native of Pennsylvania, was in the Battle of Point Pleasant and was one of the first to locate here when the new town was laid off. He purchased the grounds upon which the Mansion House in Tue Endie Wei Park now stands and built the first hewn log house in the county, which, for its beauty and size, was called the Mansion House. The house was built in 1796. In August, 1804, he was licensed to establish a ferry across both the rivers, Ohio and Kanawha, and granted a license to sell spirituous liquors and keep an ordinary at his house in the town of Point Pleasant. This was the first place in which liquor was ever legally sold in the county. Mr. Newman was also the first man in the county to support a missionary, sending his nephew, Rev. James Newman, as a missionary to South America

WILLIAM MOORE

William Moore, of Rockbridge County, after the Battle of Point Pleasant, became Capt. Wm. Moore of the Revolution. He was a merchant of Lexington, Virginia, and purchased the first bag of coffee ever purchased in Virginia, which be found slow sale for. Later, he built an iron furnace at South River in Rockbridge Co. He was a Justice of the Peace for many years and served as high sheriff of Rockbridge for two terms. He died in 1841, aged 93 years. His wife was Miss Nancy McClung, by whom he had children, Samuel, David, John, Eliah, Joseph. Isabella, Elizabeth and Nancy.

JOHN LYLE

John Lyle, of Capt. McDowell's Company, became Rev. John Lyle, of Hampshire County, now West Virginia, a pioneer minister, who proved to be a power for good in that region.

WILLIAM ROBERTSON

William Robertson of Augusta County, was commissioned a Lieutenant July, 1775. He distinguished himself at Great Bridge, Brandywine and Germantown. He died Nov. 12, 1831.

JOHN LEWIS

Captain John Lewis, son of Thomas Lewis, of Augusta Co., was with his uncle, Andrew Lewis, at Point Pleasant, where he was wounded. He was born in 1749, died 1788, leaving four children. He served under Washington at Valley Forge and in the Jerseys and witnessed the surrender of Cornwallis.

JOHN FROGG

The Sutler of the Army, was killed at Point Pleasant. Oct. 10 1774, by the Indians and was there buried. He came to Staunton from the Rappahannock and married, Miss Agatha Lewis, a daughter of Thomas Lewis a brother of General Andrew and Col. Charles Lewis, and when the Army started for the Ohio river, Mr. Frogg was appointed the Sutler and went with them.

He was a handsome young man, gallant, generous and fond of display and spoken of as a very worthy gentleman and popular with the men, and by one writer, when giving a list of the dead, spoke of him, as "poor 'John Frogg."

When he went into the battle, he had on a brilliant red jacket, which made him a prominent mark for Indians and when he fell, there were no less than five Indians that had made an attempt to secure his scalp, and all five of them were found dead on the ground where poor John lay. It is tradition that the little daughter was awakened from her sleep at three several times by the dream of her father being killed by the Indians, which she related to her mother several days before it was known that there had been a battle. Mr. Frogg was related to the Strother family, one of whom was the wife of John Madison, Clerk of Augusta Co., one was the wife of Gabriel Jones, the Crown's Attorney for said County and the other was the wife of Thomas Lewis, the surveyor of said county.

His widow became the wife of Capt. John Stuart of Greenbrier who was also in said battle; and his daughter married a Mr. Estill.

It may not have been the duty of Mr. Frogg to go into the battle at all, but it is certain he was not required to attire himself in a brilliant red jacket and make of himself a mark for sharpshooters and loose his life, but, it required five Indians to pay for his life.

Agatha Lewis, his wife, was born May 18, 1753 and she married Capt. John Stuart Nov. 18, 1776.

WILLIAM MCCORKLE

William McCorkle, who engaged in the Battle of Point Pleasant, was the son of Alexander MeCorcle (McCorkle) who had his American origin in Pennsylvania, in the Scotch-Irish reservoir of the Cumberland Valley, among the other Scotch Irish, who, emigrating to the Valleys of the Shenandoah and James, became the very seat of culture and the greatest factors in Virginia's power, and gave that state her prominence in the sisterhood of states.

In the spring of 1774 Wm. McCorkle was making preparations to emigrate with a great body of Virginians to Kentucky and, on June 3rd of that year, a survey of 1,000 acres of land was set aside for him near the present city of Louisville.

Indian hostilities necessitating the protection of the frontier, and although not a young man, Wm. McCorkle enlisted as a volunteer in Captain John Murry's Company from Botetourt and engaged in the Point Pleasant Campaign. He returned to Kockbridge County to the lands be held near Lexington, and which had continued in the possession of his descendants until 1894, when it passed into the hands of strangers.

Soon after the Battle of Point Pleasant, Wm. McCorkle died but he had offered upon his country's altar his son, John McCorkle, who, when but twenty-three years of age, was killed at the Battle of Cowpens, while serving under Gen. Morgan.

John McCorkle married Rebecca Nutt, and was survived by his two sons, Alexander and Samuel, the younger Samuel being the progenitor of five sons, the youngest of whom was William McCorkle, whose oldest son is Ex-Governor William A. McCorkle, who served as Governor of West Virginia, from March 4, 1893, to March 4, 1897, and is now located at Charleston, West Virginia, where, as a historical memorial, he has erected the most beautiful home in the state, "SUNRISE", on the summit of the mountain; embellished with historic stones and furnished and decorated with historic mementoes, demonstrating that his heredity has made him revere the past as it has made him bountifully prepare for the future.

ROBERT CAMPBELL

Robert Campbell, who was engaged in the Battle of Point Pleasant, was born in the Valley of Virginia in 1755. He was engaged throughout the Revolution. He displayed great bravery in his conflicts with the Cherokees. He was in command of a Regiment at King's Mountain, Oct. 7, 1780 For forty years he was a magistrate of Washington County, Virginia. In 1825, he emigrated to Tennessee with his children and there died.

JOHN CARTER

John Carter became a pioneer of Tennessee. During the Revolution, he was elected with John Sevier and Charles Robertson to the Convention that assembled at Hartford, N. C., in 1785, and framed a Constitution for the State of Franklin, which was reunited with North Carolina in 1788.

MATTHEW BRACKEN

Matthew Bracken had been a surveyor with Thomas Hanson from Virginia in the exploration and surveying expedition to Kentucky, which left Fincastle on April 7th, 1774. "Bracken Creek" in Mason county, Kentucky, commemorates his name. He returned to Kentucky in time to enlist as an ensign in the company of Capt. Robert McClenahan's Company of volunteers from Botetourt. He was killed in the Battle of Point Pleasant and his remains were buried within the forty foot reservation, now a part of Tu-Endie Wei Park, at the mouth of the Kanawha.

CAPT. JOHN LEWIS

Capt. John Lewis eldest son of Gen'l Andrew Lewis married Miss Patsy Love of Alexandria Virginia who had four children. His eldest son Andrew married Jane McClenahan of Botetourt County Virginia and they were the parents of six children who lived to maturity. John, William and Samuel locating in Kanawha County near the mouth of Scary.

John Lewis known as Coal River John was a man of great wealth and prominence and from him through his daughter Marjorie who married 1st Edward Kenna and 2nd Richard Ashbey has descended through this first marriage Hon. John E. Kenna who represented the old third district of West Virginia three times in Congress and was twice elected to the United States Senate each time being the youngest member in either branch of Congress. There are many descendants of Captain John Lewis living on the Kanawha including Mr. Kenna's family the family of Mr. Kenna's sister Mrs. Gentry, the family of Hon. Jos. Gaines, M. C., Mr. Walter Ash-by and many others.

THOMAS HACKET

Thomas Hacket (Haket) of Rich Creek Virginia was a member of Capt. Michael Woods Company at Point Pleasant, soon after which he settled at Petersburg Virginia where he continued to reside until after Waynes Treaty. Lured by the beauty and fertility of the Ohio Valley be settled at Kyger Ohio in Gallia County where he died and is buried having lived to the advanced age of 104 years.

Among other children he left a daughter Mary Ann Hacket who married Nimrod Kirk whose daughter Elizabeth (Betsey) Kirk married George Knight. From George Knight descended the distinguished Dr. A. L. Knight of Mason County, prominent farmer James Knight of Pleasant Flats, the late Samuel Knight of Marietta, Ohio, Mr. George Knight of Clifton, and Mrs. Louise Meeks of Dallas, Mrs. Rebecca Brown of Hartford, and Mrs. Susan Hogg of Point Pleasant and their descendants.

CAPT. JAMES CURRY

Capt. Jas. Curry served under Gen'l Andrew Lewis and was severely wounded in the right arm at the battle of Point Pleasant. His home at that time was near Staunton Virginia. When at the age of 22 years he enlisted as a private.

MICHAEL SEE

While at work outside the fort at Point Pleasant in a field near where James Capehart now resides, in August 1791 Michael See and Robert St. Clair were killed by Indians. Thomas Northrop and a colored boy belonging to See were captured and carried away prisoners.

Michael See who had been engaged in the battle of Point Pleasant was living with his family within the fort where the night of his death his wife gave birth to a son, Wm. See, from whom descend the Sees of Mason County West Virginia.

We are indebted to Rev. Price, of Marlinton and Rev. C. W. McDonnald, of Huttonsville, Randolph Co., descendants of Michael See who send the following gleaned from the history of Randolph county: "Michael See, of German ancestry, and using the language in his family is believed to have been born in Pennsylvania. He came from that state to South Branch, what is now Hardy County, W. Va., about 1765. His father's name was Frederick Michael See, of him but little is known except he had a son Adam, but Adam never lived in Randolph county Michael See, son of Frederick See was among the early settlers of Tygarts valley, Randolph Co., and his children intermarried with the following prominent families more than a century since.

Anthony See married Julia Leonard; Adam See married Margaret Warwick, daughter of Jacob Warwick, the pioneer of whose name appears in the Pocahontas sketches; Polly See, of Michael married George See. Her daughter Georgiana became Mrs. Capt. J. W. Marshall, a noted confederate officer and promoter of public improvement. Barbara See married William McCleary; John See married Miss Stewart, and Noah married Margaret Long.

COL. JAMES CURRY

March, 1900, Mr. W. S. Curry of Columbus, Ohio, Registrar of the Sons of the Revolution, sent us an account of his grandfather Col James Curry, above referred to. It follows:

"James Curry was closely engaged throughout the greater part of the day fighting from behind first one tree and then an- other but later in the day was shot through the right elbow. It is said he asked the surgeon who dressed his wound 'If it would hurt him to take a drink of wine?' to which the surgeon replied 'No if you take it with the Surgeon.' He remained in the garrison until recovered from his wounds.

He served in the army throughout the Revolution as Captain in the Fourth Va., Inft., participating in the battle of Brandywine, Yorktown. He was wounded at the siege of Charleston, S. C. and taken prisoner May 12, 1871 by Gen. Lincoln's Army.

After the Revolution he settled at Staunton, served a term as Clerk of the Court for Augusta Co. In 1797 he moved to Highland County Ohio where he entered a Virginia Military tract of land.

SOLOMON BRUMFIELD

Solomon Brumfield who enlisted under Gen'l. Andrew Lewis at Staunton resided where the city of Washington is now built.

WM. HAMILTON

Wm. Hamilton was an orderly at Point Pleasant in the army of Gen'l. Lewis and when the battle began was sent as a messenger to hasten the regiments in command of Capt. Wm. Christion whom he met at the mouth of 13 mile creek, the present site of the village of Leon. Here a few were left in charge of supplies while the remainder of the companies marched on to Point Pleasant arriving at about 11 o'clock.

BAZALEEL WELLS

Bazaleel Wells, afterward Gen'l Wells of the Revolution, became the founder of Steubenville, Ohio, and helped to form the first Constitution of Ohio. He was then a member of the Ohio Senate and was probably the wealthiest man in Eastern Ohio at the beginning of the century. It was he who financed the building of the Zanesville road the great highway of Ohio that proved to be the great artery of commerce of that successful, progressive new state.

JOHN MURRY, EARL OF DUNMORE (LORD DUNMORE)

In reviewing, (by many historians,) the life and character of Lord Dunmore, there are none who have more truly recorded his character than Hu Maxwell, in his *History of Hampshire County*, who says on page 51:

"Before proceeding to a narrative of the events of the Dunmore War, it is not out of place to inquire concerning Governor Dunmore, and whether from his past acts and general character he would likely conspire with the British and the Indians to destroy the western settlements in Virginia. Whether the British were capable of an act so savage and unjust as inciting savages to harrass the western frontier of their own colonies is not a matter for controversy. It is a fact that they did do it during the Revolutionary War."

As to a confirmation of the character of Dunmore and his methods we quote again from the same author, relating to events subsequent to the Battle of Point Pleasant:

"Dunmore had trouble elsewhere. His domineering conduct and his support of some of Great Britain's oppressive measures, caused him to be hated by the Virginians, and led to armed resistance. Thereupon be threatened to make Virginia a solitude using these word; "I do enjoin the true and loyal subjects to repair to my assistance, or I shall consider the whole country in Rebellion and myself at liberty to annoy it by every possible means, and I shall not hesitate to reduce houses to ashes, and spreading devastation wherever I can reach. With a small body of troops and arms, I could raise such a force from among Indians, negroes and other persons as would soon reduce refractory people of them."

The patriots finally rose in arms and drove Dunmore from the country. Some of these events occurred after the Dunmore War, but they serve to show what kind of a man the Governor was."

He was born in England in May, 1709, descending from the house of Stuart. He succeeded to the peerage in 1756; appointed Governor of New York 1770; of Virginia, July, 1771. With a band of Tory followers, he plundered the inhabitants, residing on the James and York Rivers. He and his followers suffered defeat at the battle of Great Bridge, shortly after which he burned Norfolk, then the most prosperous city in Virginia. In

1779, he returned to England, and, in 1786, was appointed Governor of the Bermudas.

LOGAN

Tah-gah-jute, son of Shikellanny, chief of the Cayuga Indians, was born in about 1725. He was named Logan by the whites for James Logan, (the Secretary of Wm. Penn,) who had been such a friend of the Indians, that they accepted the name as a badge of honor.

Logan was reared near the Moravian settlements and his relation to the whites had been most friendly. Throughout Virginia and Pennsylvania he was known for his commanding presence and engaging qualities. Residing with his family near Readsville, Penn., he supported them by hunting, and trapping and dressing hides which he disposed of to the whites. At this time, the Mingo tribe of Indians chose him as their chief. About 1770, he moved to the banks of the Ohio, where, in the spring of 1774, his family were massacred, whereupon he sent a declaration of war to Michael Cressap, whom he believed had ordered the massacre. Thus began the fearful depredations that burst upon the frontier and Logan is said to have taken thirty scalps himself before the termination of hostilities at Point Pleasant, where he was engaged that fearful October day. Subsequent to the battle when the other. Indian chiefs sued for peace, Logan disdained to participate in the treaty. Governor Dunmore sent John Gibson to personally invite him to the council and he then gave out that classic in English literature that has made so famous the name of Logan. It is as follows:

"I appeal to any white man to say if he ever entered Logan's cabin hungry, and he gave him not meat; if he ever came cold and naked and he clothed him not. During the course of the last long and bloody war, Logan remained idle in his cabin, an advocate of peace. Such was my love for the whites that my Countrymen pointed as they passed and said: 'Logan is the friend of the white men.' I had even thought to have lived with you but for the injuries of one man. Cresap, the last spring, in cold blood and unprovoked, murdered all of the relations of Logan, not even sparing my woman and children. There runs not a drop of my blood in the veins of any living creature. This called for my revenge. I have sought it; I have fully glutted my vengeance. For my country, I rejoice at the beams of peace. But do not harbor the thought that mine is the joy of fear; Logan never felt fear. He will not turn on his heel to save his life. Who is there to mourn for Logan? Not one."

The speech was written down, when Gibson repeated it to an officer, and published in the *Virginia Gazette.* Thomas Jefferson, a great admirer of Logan, took pains to establish its authenticity and published it in his *Notes on Virginia.*

Logan was killed by his nephew at Lake Erie in the summer of 1780.

CORNSTALK

Cornstalk, the celebrated Shawnee warrior, is first mentioned in Colonial History in 1763, when about sixty Indians, led by Cornstalk attacked the settlement on Muddy Creek, in Greenbrier County, Virginia, when they pretended to be on a friendly mission, at which time they arose and murdered all except a few women and children, whom they took prisoners. From there the Indians went on to the Levels in the same county, to the home of Alexander Clendennin, where many were gathered enjoying the fruits of a successful chase and the Indians, too were treated with the utmost hospitality, but they again murdered most of the inmates of that place. Mrs. Clendennin was carried away a prisoner and with others taken to Muddy Creek.

For a year the Indian depredations were continued, until there was not a white settlers left in Greenbrier County which was not again inhabited by whites until 1769, when Col. John Stuart and a few others became permanent settlers.

It is said that Cornstalk was born in the Kanawha Valley about 1727.

In the Battle of Point Pleasant, he commanded the army consisting of the flower of the Shawnee, Delaware, Wyandotte, Mingo and Cayuga braves, he being the King of the federation, in their herculian efforts to stay the oncoming tide of Saxon civilization. These Indians were, fighting to maintain their homes and their hunting grounds, and, if the whites were eyed to be repelled, it must be now.

This was not the first time in battle array that the Shawnees had shown their skill as warriors. In the Braddock defeat and other campaigns they had proven themselves valiant. They despised treaties and had chafed under that with Boquet so that at the Battle of Point Pleasant, they had determined to be victorious. It was not that they favored Great Briton. All whites were alike to them except as they availed to help them save their hunting grounds; and here were gathered their ablest leaders: Cornstalk, Red Eagle, Scoppathus, Blue Jacket, Logan, Chief of the Cayngas, Illinipsico, Red Hawk, the noted Delaware Chief and others commanding the most formidable army every arrayed as an Indian phalanx. The story of their bravery has already been related in the accounts of the Battle of Point Pleasant, here in published.

It were well for the white settlers, if the history of Cornstalk could have ended with the Battle at Point Pleasant. The treaty with the whites following the battle, was maintained in the highest sense of honor by Cornstalk.

In the spring of 1777, when the great Indian uprising was again taking place, Cornstalk came to Fort Randolph at Point Pleasant to warn the whites of their danger and was retained as a hostage, during the whole of the summer. In November, his son, Illinipsico, came in search of his father, hallooing to be brought across the river. The next day two hunters crossed the Kanawha and, returning, one was killed by Indians. Some of the whites made a rush for the Indians detained at the fort. Captain Arbuckle in command tried to stay them, but incited by one of the Gilmores whose kindred had been massacred at Muddy Creek, Corn- stalk and his son, Illinipsico and Red Hawk were murdered in cold blood, by Captain James Hall and Hugh Galbraith leading the men. The bravery of Cornstalk called forth the admiration of even his brutal murderers, as he thus addressed Illinipsico.

"My Son, the Great Spirit has seen fit that we should die together, and has sent you here to that end. It is His will and let us submit; it is all for the best!" and then turning his face to his murderers at the door, he fell without a groan pierced with seven bullets."

Cornstalk said the day before he was killed while attending a conference with the whites. "When I was a young man and went to war, I often thought each might be my last adventure, and I should return no more. I still lived. Now I am in the midst of you and if you choose you may kill me. I can die but once. It is alike to me whether now or hereafter."

From the records of Rockbridge County we quote the following.

"At a court held in Rockbridge County, April 18, 1778, for the examination of Capt. James Hall and Hugh Gailbraith, charged with the murder of Cornstalk, no witnesses appearing, they were acquitted for the murder of Cornstalk and two other Indians in November, 1777 they denying their guilt."

The remains of Cornstalk were interred at Point Pleasant outside the fort, where Viand intersects with First Street, but in August 4, 1840, when Viand Street was opened, his remains were removed to the Court House yard. Dr. Samuel G. Shaw made a memorandum at the time of his burial. His grave is thirty yards in the rear of the Court House where the grave is neglected.

On October 13, 1899, a monument to Cornstalk with the simple inscription

"CORNSTALK"

was erected in the Court House yard near Sixth Street. The monument is of grey limestone the stone for which was donated by Mr. S. H. Reynolds, then Superintendent of Construction at Kanawba Lock 11. The money for the completion of the monument was raised by private subscription and the dedication of the monument was the occasion for a public ceremonial with a dedicatory address by Hon. C. E. Hogg, Mayor John E. Beller receiving the monument for the town.

FORTS BLAIR, RANDOLPH AND POINT PLEASANT

Governor Dunmore under date of June 12, 1774, directed Gen'l. Andrew Lewis "to collect a body of men immediately; go down to the mouth of the Great Kanaway and THERE BUILD A FORT; and then if you have force enough to invade the Indian Country, do so."

Before these orders could be carried out the battle of Point Pleasant had ensued. The wounded of the army must be cared for, and sufficient of the army must remain to protect and care for them while the majority marched on with Gen'l. Lewis to Camp Charlotte.

So frail was the hastily improvised stockade that it was inadequate to withstand an attack but fortunately Capt. Wm. Russell and fifty Fincastle men were delegated by General Lewis to return to Point Pleasant and erect a fort. They arrived there November 11, 1774. Thus Captain Russell became the designer and builder of this small rectangular pallisade, eighty yards long with block houses at two corners with cabins for barracks, which he called Fort Blair. It was erected upon grounds on the North Bank of the Kanawha as it emptied into the Ohio. Here in January, 1775, Cornstalk complying with the terms of the treaty at Camp Charlotte delivered a number of white prisoners.

On June 5, 1775, Governor Dunmore reported that the garrison at Point Pleasant had been ordered discontinued, but the Virginia convention ordered that one hundred men should be hastened to Point Pleasant. Upon their arrival they found Fort Blair in ashes.

By whom or when destroyed history does not record. On May, 16, 1776, Captain Matthew Arbuckle passed through Wheeling enroute to Point Pleasant where upon the ruins of Fort Blair he erected Fort Randolph. This fort was much larger than Fort Blair, called Fort Randolph in honor of Hon. Peyton Randolph of Virginia. It was from thence garrisoned at expense of the colony of Virginia Captain Arbuckle continuing in command until the close of 1777. He was succeeded in command by Capt. Wm. McKee of Rockbridge County. In 1778 several were killed by Indians while outside the fort at work, including Lieut. Moore. In 1779, for a week Indians besieged the fort but to no avail except that they captured all the cattle. In 1779 prior to July 12, Ft. Randolph was evacuated after which it was burned by Indians. Capt. Andrew Lewis visiting at Point Pleasant in 1784 reported. There was then but little or no sign of the fort to be seen. In 1785 a third fort was built at Point Pleasant, on the Ohio River above the present First street. Commanded by Colonel Thomas Lewis and from that year on the white man has never ceased to reside at Point Pleasant.

PARTICIPANTS OF THE BATTLE.

No official roster having been preserved either by the Government or State, the following list has been gleaned from the sources available after years of research by the writer.

Adams, John

Adkins, Parker

Adkins, Wilton

Agnew, (Aggnue), John

Albe, Jeremiah

Alden, Andrew

Alexander, James Ser.

Allen, Hugh Lieut.

Allen, James

Allen, James

Allen, Thomas

Alley, Thomas

Alliet (Elliot) Robert

Alsbury, Thomas

Anderson, James

Anderson, Samuel

Andrews, Samuel

Arbuckle, Capt. Matthew

Arbuckle, John

Ard, James

Armstrong Thos.

Armstrong, Geo.

Armstrong, Wm.

Arnold, James

Arnold, Steven

Arthur, John

Astle, Samuel
Atkins, Blackburn
Atkins, Charles
Atkins, Henry
Atkins, Wm.
Babbit, Ishmael
Bailey, Campbell
Baily, John
Baker, Ensign Samuel
Baker, Markham
Baker, Martin
Baker, Thomas
Ball, James
Bambridge, James
Baret, Edward
Barker, Samuel
Barkly, John
Barnes, ---
Barnett, James
Barnett, S. L.
Barton, Samuel
Basel John
Bates, ---
Bazel, John
Bell, Thomas
Bellew, Daniel (Canoe man)
Bergman, Christian
Berry, Francis
Bishop, Levi
Blackburn, Arthur
Blackford, Joseph

Blair, Daniel

Blair, Wm.

Blankenship, Richard

Bledsoe, Abraham Lieut.

Blesly, Jacob

Blesly, John

Boh, Adam

Boh, Jacob

Bojard, Abraham

Boles, John

Boniface, Wm.

Borg, Francis

Boughman, Jacobs

Boughman, John

Bowen, Moses

Bowen, Reese

Bowen, Wm.

Bowles, --- Sergt.

Bowles, Robt.

Bowyer, Henry

Boyd, James

Boyd, Robert

Boyer, Henry

Boyles, Barney

Boylstone, ---

Boylstone, Wm.

Bracken, Matthews Ensign & Lieut.

Bradley, John

Bradley, John

Bradley, Wm.

Brambradge, Jas.

Bramstead, Andrew

Breckinridge, Alexander

Breden, John

Breeze, Richard

Breeze, Robt.

Brooks Geo.

Brooks, Thos.

Brown, Chas.

Brown, James

Brown, Low

Brown, Robt.

Brown, Thos.

Brown, Wm.

Brumfield, Humphrey

Brumfield, Solomon

Brumley, Thos.

Bryans, Shorgan

Bryans, Wm. Sergt.

Bryant, Wm.

Buch, Sergt. John

Buch, Wm.

Buchanan Col. John

Buchanan, --- (Commissariat)

Buchanan, Ensign Wm.

Buchnell, John

Buford, Col. Abraham

Bunch, Joseph

Burch, Richard

Burcks, Samuel

Burk, John

Burk, Thos.

Burnes, Thos.

Burney, Thomas

Burnsides, James

Burrens, Jarnes

Burroughs John

Burtchfield, James

Burton, Litton

Buster, David

Butler, Joseph

Butlor, Shabrick

Byrd, Richard

Byrne, Chas

Calloway, Dudley

Calron, Michael

Cameron, Geo.

Cameron, Hugh

Campbell Joseph

Campbell, Arthur Maj.

Campbell, John Capt.

Campbell, Robt.

Campbell, Samuel

Campbell, Wm. Capt.

Canady, Thos.

Caperton, Adam

Caperton, Hugh

Cariain, Joel

Carlton, James

Carmack, John

Carney, Martin

Carpenter Solomon

Carpenter, Jeremiah

Carpenter, John
Carpenter, Thomas
Carr, Geo.
Carr, John
Carr, Wm.
Cartain, James
Cartain, John
Carter, John
Carther, Edward
Cary, Jeremiah
Casey, Wm.
Cashaday, Thos.
Cashady, Simon
Catron, Adam
Catron, Francis
Catron, Jacob
Catron, Peter
Catron, Philip
Cats, Roger
Cattes, John
Cavenaugh Wm.
Cavenaugh, Charles
Cavenaugh, Philemon
Cavenaugh, Philip
Cecil, Saul
Champ, Wm.
Chapline, Abraham
Chapman, John
Chapman, Richard
Charlton James
Chesney, John

Christian, Col Wm.

Clark, James

Clark, John

Clark, Samuel

Clay, David

Clay, Mitchell

Clay, Zekel

Clendinen Adam

Clendinen Chas.

Clendinen Geo.

Clendinen, Alexander

Clendinen, Robert

Clendinen, Wm.

Clerk, John

Clifton, Wm.

Clinding, Geo.

Clinding, Wm.

Cloyne Nicholas

Cochran, Wm.

Cocke, Capt. Wm.

Coile, James

Coller, John

Coller, Moses Sergt.

Collet, Thos.

Collins, Richard

Condon, David (canoe man)

Conner, Patrick

Conner, Wm.

Constantine Patrick

Cook, David

Cook, Henry

Cook, John

Cook, Wm.

Cooper, Abraham

Cooper, Francis

Cooper, Leonard

Cooper, Nathiel

Cooper, Spencer

Cooper, Thomas

Copley, Thos.

Corder, John

Cormick, John

Cornwell, Adam

Cornwell, Adam

Cornwell, John

Courtney, Chas.

Courtney, John

Cowan, Jared

Cowan, John

Coward, ---

Cox, Capt. John

Cox, Lieut. Gabriel

Coyl, James

Crabtree, Wm. (scout)

Craig, George

Craig, John

Craig, Wm. Serg.

Crain, John

Craven, Joseph Serg.

Cravens, James

Cravens, John

Cravens, Robt.

Crawford, Bonard

Crawford, John Serg.

Crawley; (Croley) James

Creed, Matthew

Crisman, Isaac

Crockett, Capt. Walter

Crockett, Joseph

Croley, Samuel

Crow, John Serg.

Crow, Wm.

Culwell, Alexander

Cummins, Geo.

Cundiff, Johnathan Ensign

Cunningham James

Cunningham John

Current, Joseph

Curry, James, Capt.

Curwell, Alexander

Custer, Wm.

Cutlep, David

Cutright, John

Dale, James

Daugherty, Geo. Serg.

Daugherty, James

Daugherty, John

Daugherty, Michael Serg.

Davis, Charles

Davis, Geo.

Davis, Robert (scout)

Davis, Samuel

Davis. Capt. Azariah

Davise, Johnathen

Day, Joseph

Day, Wm.

Deal, Wm.

Deek, John

Demonse, Abraham

Deniston, John

Denton, John

Dickinson, Col. John

Dillon, --- Lieut.

Dingos, Peter

Divey, John

Doack, David

Doack, Robt. Capt.

Doack, Samuel

Doack, Wm. Ensign

Dobler, Jacob

Dodd, James

Dodd, John

Dollarhide, Samuel

Donaley, John (fifer)

Donaley, Serg. James

Donalson, Col. John

Donalson, Robt.

Donalson, Thos.

Donley, Jacob

Dooley, Thos. Lieut.

Doran, Patrick

Dorherty, James

Dorherty, John

Doss, Joel

Douglas, Geo.

Douglas, James

Downy, James Serg.

Downy, John

Drake, Ephriam

Drake, Joseph

Draper, Lieut. John

Dulin, James

Duncan, John Serg.

Dunkirk, John Serg.

Dunlap, Robert

Dunn, John

Dunowho, James

Duttan, Philip

Dyer, Wm.

Eager, John

Eastham, Geo.

Eastham, Wm.

Edgar, Thos.

Edmiston (Edmondson) Wm. Lieut.

Edward, James

Edwards, Johnathan

Egnis, Edward

Elias, Thomas

Elkins, Jesse

Ellenborough, Peter

Elliott, Capt. Robert

Ellison, Charles

Ellison, James

Elswick. John

English, Joseph

English, Joshua
English, Stephen
Estill, Samuel
Evans, Andrew
Evans, Evan
Ewing Samuel
Ewing, --- Jr.
Ewing, Alexander
Ewing, Robert
Ewing, Wm. Serg. & Maj.
Fain, John
Fain, Samuel
Fargison, Thos.
Farley (Farlen), Francis
Farley (Farlor), Thomas
Farley, --- John
Farmer, Nathan
Feavil, Wm.
Fendlay, Robt. Serg.
Fenquay, Isham (canoe man)
Ferrill, Robt.
Ferrill, Wm.
Field, Col. John
Fielder, John Serg.
Fielder, Wm.
Fields, Wm.
Findlay, Geo.
Findlay, John
Fisher, Isaac
Fitspatrick, Timothy
Fitzhugh, John

Fleming, Col Wm.

Flintham, John

Fliping, Thos.

Floyd, John

Fourgeson, Thos.

Fowler, Jas. (scout)

Fowler, Samuel

Fowler, Wm.

Franklin, James

Franklin, Wm.

Frazer, John

Freeland, John

Friel, Jeremiah

Frogg, Lieut. John

Fry, Geo.

Fry, Geo. Jr.

Fry, John

Fullen, Chas.

Fullen, Daniel

Fullen, James

Fuls, Geo.

Gardner, Andrew

Garrett, Wm.

Gass, David (Messenger)

Gatliff (Gatlepp), Squire

Gibbs, Luman

Gibson, Joseph

Gilberts, Thos.

Gilkenson, Jas.

Gill, Prisley

Gillass, Wm.

Gillespy, Thos.

Gillihan (Gilliland), John

Gillman, Duncan

Gilmore, Capt. Jas.

Gilmore, John

Givens, Lieut. ---

Glascum, David

Glass Wm.

Glass, Serg. Samuel

Glaves, Michael

Glenn, Davis

Goff, Andrew

Goldman, Lieut. Edward

Goldsby, ---

Goodall, John

Gordan, Moses

Gorman, David

Graham, Benj.

Green, John

Griffin, Robt.

Grigger, Michael

Grigger, Peter

Grigs, John

Grigsby, ---

Grimp, John

Guffy, James

Guillen, Edward James

Gullion, Barney

Gurden, Michael

Hackett, Thos.

Hackworth Wm.

Hackworth, Augustine

Haines, Lewis

Hale, Edward

Hale, Thomas

Hale, Wm

Hall, James

Hall, Thos.

Hamilton, Francis

Hamilton, Isaiah

Hamilton, Jacob

Hamilton, Jas.

Hamilton, John

Hamilton, Thos.

Hammond, Philip

Hamrick, Thos.

Hamrick, Wm.

Handley (Hensley), Sam'l

Handley (Herrill), Robt.

Hanee, Philip

Hansburger, Adam

Hanson, Wm.

Harlan, Elijah

Harlan, Silas

Harmon, Dangerfield

Harmon, Geo.

Harmon, Israel

Harmon, John

Harrel, Wm. (scout)

Harriman, Skid Serg.

Harris Griffin

Harris, John

Harris, Stephen
Harrison, Andrew
Harrison, Benj. Capt.
Harrison, John Lieut.
Harrod, James Capt.
Hart, Thos.
Hasket, Thos.
Hatfield, Andrew
Haugh, Jacob
Havens James
Havens, John
Hayes, John
Haynes, Benj.
Haynes, Capt. Joseph
Hays, Chas.
Hays, Samuel (scout)
Hckman, ---
Head, Anthony (Messenger)
Hedden, Thos.
Hedrick, Peter
Henderson, Alexander
Henderson, Daniel
Henderson, Lieut. John
Henderson, Sam'l
Hendrix, Peter
Henly, Geo.
Henly, Wm.
Hensley (Hadley), Sam'l
Hepenstahl (Hempinstall), Abraham
Herbert, Wm. Capt.
Herd, Richard

Herrill, Robt.

Higgans (Higans,) Peter

Higgans, Philemon

Hill, --- Capt.

Hill, James

Hill, Robert

Hobbs, Vincent

Hogan, Henry

Hogan, Win.

Holley, Wm.

Hollway (Holloway), Richard

Holston, Stephen

Holwell, Walter

Homes, Lewis

Hooper, Wm.

Hopton, Stephen

Hopton, Wm.

Horne, Joseph

Howard, Charles

Howard, Henry

Huchisen (Hutchinson), Wm.

Huff, Leonard

Huff, Peter

Huff, Samuel

Huff, Thomas

Hughes, Davy

Hughes, Ellis

Hughey, Joseph

Humphries, John

Hundley, John

Hunter, Robert

Hutchinson, Lewis

Hutson, John

Hynes, Frances

Ingles, Wm. Major (Commissary)

Inglis, Joshua

Inglis, Thos.

Inglish (English) Joshua

Inglish (English), Joseph

Irvine, John

Isum, Wm.

Jackson, Yerty

Jameson, John

Jenkins, Jeremiah

Jennings, Edmond

Jewitt, Matthew

Johns, Wm.

Johnson, Capt. Arthur

Johnston, John

Johnston, Patrick

Jones, Benj.

Jones, John

Jones, Thos.

Jones, Wm.

Kasheday, Peter

Katcliff, Wm.

Keeneson, Charles

Keith, Samuel

Kelley, Alexander

Kelsey, John

Kendrick, James

Kennedy, Ezekiel

Kennedy, Thomas

Kennedy, Wm. Serg.

Kenneson, Chas.

Kenneson, Edward

Kennot, Zacariah

Kerr, James

Kerr, Wm

Kincaid (Kingkeid), David

Kincaid (Kingkeid), David Jr.

Kincaid (Kingkeid), Geo.

Kincaid (Kingkeid), James

Kincaid (Kingkeid), John Serg.

Kinder, George

Kinder, Jacob

Kinder, Peter

King, James

King, John

King, Wm.

Kinsor, Michael

Kinsor, Chas.

Kinsor, Jacob

Kinsor, Walter

Kishoner, Andrew Jr.

Kishoner, Andrew Sr.

Kissinger, Andrew

Kissinger, Matthew

Knox, James

Lammey, Andrew

Lapsley, John

Larken, John Serg.

Lashly, John

Laughlin, James
Lawrence, Henry
Learned (Lord), Lieut.
Lee, Sefinah
Lee, Zacarias
Lemaster, Richard
Lesley, John
Lesley, Wm.
Lesley, Wm. Adj.
Lester, John
Lester, Samuel
Lewis, Andrew Gen.
Lewis, Benjamin
Lewis, Chas. Col.
Lewis, John Capt. of Augusta
Lewis, John Capt. of Botetourt
Lewis, John Private
Lewis, Samuel
Librough, Henry
Lin, Adam
Litton, Burton
Litton, Solomon
Litz, Wm.
Lockhart, Jacob
Lockhart, Queavy
Lockridge, Andrew Capt.
Logan, Benj.
Logan, Hugh
Logan, James
Logan, John
Long, Joseph Ensign

Lord (Leord), Lieut.

Love, Joseph

Love, Philip Capt.

Luallen, Thos.

Lucas, Chas.

Lucas, Chas. Jr.

Lucas, Henry

Lucas, John

Lucas, Wm.

Luney, Michael

Lybrook (Librough), Henry

Lybrook (Librough), Palser

Lyle, John

Lyman, Richard

Lyn, James

Lynch, ---

Lyons, Wm.

Madison, John

Manadue, Henry

Mann, John

Mann, Wm.

Marks, John

Martin, Brice

Martin, Christian

Martin, Geo. Jr.

Martin, Geo. Sr.

Martin, Philip

Martin, Wm. Col.

Matthew, Capt. Geo.

Matthew, Sampson

Maxwell, Bezaleel

Maxwell, David

Maxwell, John

Maxwell, Thomas

Mayse, Joseph

McAfee, Geo.

McAfee, James

McAfee, Robt.

McAfee, Samuel

McAfee, Wm

McAllister, Wm.

McBride, James

McBride, Joseph

McCallister, James

McCallister, Wm.

McCandless, John

McCartney, John

McCarty, James

McCastem, Wm.

McClanaham, John, (Canoe man)

McClanahan, Absalom

McClanahan, Alexander, Capt.

McClanahan, Robt. Capt.

McClintic, Wm.

McClure, John

McClure, Thomas

McCorkle, Wm.

McCoy, Wrn. Lieut

McCune, Wm.

McCutchen, Wm.

McDonald, Daniel

McDonald, James

McDowell, Archibald

McDowell, M. Capt.

McDowell, Samuel Capt.

McElhaney, Francis

McFarland, Robt.

McFarland, Wm.

McGee, John

McGeehey, Samuel

McGinness, John

McGlahlen, John

McGuff, John

McGuff, Patrick

McKee, Wm. Capt.

McKinnett, Alex

McKinney, John

McKinsey, Hensley

McKinsey, Moredock

McLaughlin, Edward

McMullen, Wm.

McMullin, John

McNeal (Niel), John

McNiel, Peter

McNiell, Daniel Lieut.

McNutt, Alexander

McNutt, James

Mead, Nicholas

Mead, Thos.

Meader, Israel

Mecrary, Thomas

Meek, Wm.

Mercer, Hugh

Messersnuth, Barnett

Messersnuth, John

Micalister, Wm

Milican, John

Miller, James

Miller, Robert

Mills, John

Milwood, Geo.

Miner, Henry

Mitchell, James

Mitchell, James Capt.

Mitchell, Thos.

Moffat (Manford), Robt. Capt.

Moffat, George Capt.

Montgomery, Jas. Capt.

Montgomery, Samuel

Moody, John

Moon, Abraham

Mooney, Frederick

Mooney, Hendly Ensign

Mooney, Hugh

Mooney, James

Mooney, John

Mooney, Moses

Mooney, Nicholas

Mooney, Samuel

Mooney, Wm.

Moor, Frederick

Moor, Hugh

Moor, John

Moor, Moses

Moor, Wm.

Morris, Wm.

Morrow, James

Morrow, James Jr.

Mullin, Thos.

Mungle (Mongle), Daniel

Mungle (Mongle), Frederick

Murry, James

Murry, John Capt.

Myers, Wm.

Nail, Dennis

Nail, Thomas

Nail, Thomas

Nalle, Martin Lieut.

Naul (Nowl), Wm. Capt.

Nave, Conrad

Neal, Wm.

Neaville, John

Neaville, Joseph

Neely, James (Cadet)

Neely, Wm.

Neil, John

Nelson, John

Newberry, Joseph

Newell, James

Newland, Abraham

Newland, Isaac

Newland, John

Newman, Walter

Nicholas, John

Nickels, Isaac

Noland, John

Nowell, John

Null, Jacob

Null, John

Odear, James

Oguillen, Barnett

Oguillen, Duncan

Oguillen, Hugh

Oguillen, John

O'Haara, Chas. Capt.

O'Haara, Robt.

O'Haara, Wm.

Oharron, Henry

Olverson, Joseph

Ormsbey, Daniel

Overstreet, Wm.

Owen, Robt.

Owens, David

Ower, Thomas

Owler, Henry

Owler, John

Pack, George

Pack, Samuel

Packwood, Richard

Pain, Joseph

Parchment, Peter

Parsons, James

Pate Jeremiah

Patten, John

Pauling, Henry Capt.

Paulley, James

Paulley, John

Pawlings, Moses

Paxton, Samuel

Peary, Thomas

Pence, Jacob Ensign

Perce, Thomas

Peregin, Molastin

Persinger, Jacob

Petty, Benjamin

Peyton, John

Peyton, Rowzie

Pharis, Wm.

Pierce, John

Pierce, Lieut

Plunkenpel, Zacarias

Poage, Wm. Serg.

Poling, Mathew

Portor, Robt.

Posey, Thos. (Commissary)

Potter, Thos.

Preston, Wm.

Price, James

Price, Reese

Price, Rickard

Price, Thomas

Price, Wm.

Pricket (Pucket), Drury

Priest, David

Priest, Samuel

Priest, Wm.

Pright, John

Prince, Wm.

Prior (Pryor), John

Rains, Robt.

Ramsey, Josiah

Ranis, Robert

Rapp, Frederick

Ratcliff, Matthew

Ravenscroft, Thos.

Ray, Wm.

Razor, Michael

Read, John Ensign

Reagh, Archibald

Reagh, John

Reary, James

Reburn, John

Rediford, Benj.

Reed, Alexander

Reese, Andrew

Reid, Andrew

Reid, Thos.

Reynolds, John

Richardson, Benj.

Richardson, Wm.

Riley, John

Roay, Joseph

Roay, Joseph

Roberts, John

Robertson, James

Robertson, Jas. Capt.

Robertson, Thos. Major

Robertson, Wm.

Robertson, Wm. Lieut.

Robinson, Elijah

Robinson, Hugh

Robinson, John

Robison, Isaac

Robison, Jas. Lieut.

Robison, Julius

Robison, Wm.

Roe, --- Capt.

Rogers (Rodgers), Andrew

Rogers (Rodgers), Chesley

Rogers (Rodgers), David

Rogers (Rodgers), James

Rogers (Rodgers), Thos.

Rogers (Rodgers), Wm.

Rollens, Richard

Ross, Edward

Ross, Tavener

Rowan, Francis

Rucker, Geo.

Ruddle (Riddle), Geo.

Rue, Abraham

Russell, Wm. Gen.

Rutheford, Benj.

Salsbury, Wm.

Samples, Samuel

Sanders, James

Sappington, Daniel

Savage, John

Savage, Samuel

Sawyers, John Col.

Sayres, John

Scails, Wm.

Scarbara, James

Scard, --- Lieut.

Scott, Archelaus

Scott, Archibald

Scott, Daniel Capt.

Scott, Geo.

Scott, James

Scott, Wm.

Sedbery, John

See, Michael

Seed, Francis

Selby, James

Sevier, John Gen.

Sevier, Valentine

Shain, John

Shannon, John

Shannon, Samuel

Sharp, Abraham

Sharp, Edward

Sharp, John (Scout)

Shaw, Henry

Shelby, Evan Capt.

Shelby, Isaac Lieut.

Shelby, James

Shelby, Wm. Capt.

Shell, Arnold

Shelp, John

Shillin, John

Shoatt, Emanuel

Simmerman, Geo.

Simms, Chas.

Simpkins, Daniel

Simpkins, James

Simpson, James

Simpson, John

Simpson, Wm.

Skaggs, Reuben

Skaggs, Zach

Skidmore, John Capt.

Slaughter, --- Capt.

Slaughter, Francis Col.

Slaughter, Geo. Col.

Slaughter, Lawrence

Smith, Bruten

Smith, Daniel Capt.

Smith, David

Smith, Edward

Smith, Ericus

Smith, James

Smith, John

Smith, Mecagh

Smith, Moses

Smith, Robt.

Smith, Wm.

Smithers, Gabriel

Sobe, Geo.

Spicer, Wm.

Spratt, Isaac Serg.

Squires, Uriah

Staffy, Michael

Stailey, Martin

Steele, Andrew

Steele, John

Stephens, John Lieut.

Stephens, Stephen

Stephens, Thomas

Stephens, Wm.

Stephenson, Hugh Capt.

Stephenson, Robt.

Sterns, Conrad

Stevens, ---

Steward, John

Steward, Walter

Stewart, John

Stewart, John Capt.

Stewart, Wm.

Stull, Martin

Stump, Michael

Sullivan, James

Sullivan, Sam'l

Summers, Charles

Swoop, John

Tarney (Farney), Peter

Tate, T. Lieut.

Tate, Wm.

Taylor, --- Capt.

Taylor, Daniel

Taylor, Isaac

Taylor, Seiltor

Taylor, Wm.

Teasy, Wm.

Terrence (Torrence), Andrw
Thomas, Edward
Thompson, Andrew Ensign
Thompson, Richard
Thompson, Robert
Thompson, Wm.
Tipton, John
Todd, James
Todd, John
Trent, --- (Canoe Master)
Trent, Obediah
Trimble, Isaac
Trimble, James
Trotter, John
Trotter, Richard
Tucker, Wm.
Tyler, Isaac
Vails, John
Vallendingham, Geo.
VanBibber, Isaac
VanBibber, Jesse
VanBibber, John
VanBibber, Mathias
VanBibber, Peter
Vanbrook, Samuel
Vance, Edward
Vance, Samuel Lieut.
Vanhook, Samuel
Vaughan (Vaun), John
Vaut (Vaught), Andrew
Vaut (Vaught), Christian

Vaut (Vaught), Geo.

Venable, Wm.

Waggoner, Andrew

Waggoner, Henry

Waggoner, Henry Jr.

Walker, Adam

Walker, Henry

Walker, James

Wallace, Adam Ensign

Wallace, Andrew

Wallace, David

Wallace, Robt.

Wallace, Samuel Lieut.

Walter, Michael

Wambler, Geo.

Wambler, Mitchell

Ward, David Ensign

Ward, James Capt.

Ward, Wm. Serg.

Warwick, Jacob

Washburn, James

Washburn, Steven

Watkins, Robt.

Watson, Johnathan,

Waugh, --- Cadet

Weaver, Christian

Weaver, Michael

Welch, James

Welch, John

Welch, Thomas

Welch, Thomas Jr.

Wells, Bazaleel

Wells, Samuel

Welsh, Christopher

Wetzel, John

Wetzel, Martin

Whish, Richard

White, Davi

White, Joseph

White, Solomon

White, Wm.

Whitley, Moses

Whitticor, Joseph

Whitton, Jerremiah

Whitton, Thomas

Whitton, Thomas Jr.

Wiles, Robert

Wiley, James

Wiley, Robert Jr.

Wiley, Thomas

Williams, Alden

Williams, David

Williams, Isaac

Williams, James

Williams, Jarrett

Williams, John

Williams, Mack

Williams, Philip

Williams, Richard

Williams, Rowland

Williams, Samuel

Williams, Thomas Serg.

Williamson, Aldin
Williamson, David
Willis, Henry
Wilmoth, Wm.
Wilson, Benj.
Wilson, Edward
Wilson, George
Wilson, James
Wilson, John Capt.
Wilson, Samuel Capt.
Wilson, Thomas
Wilson, Wm. Serg.
Wood, Adam
Wood, Andrew
Wood, Archibald
Wood, James Serg.
Wood, John
Wood, Jos. Capt.
Wood, Michael Capt.
Wood, Richard
Woodburn, James
Woodburn, Steven
Woolsey, Richard
Workman, Daniel
Young, John

HISTORY OF THE MONUMENT BUILDING

In our research for information relative to efforts being made to erect a battle monument at Point Pleasant, the earliest record of which we have an account is a letter yet preserved, written by Hon. J. M. H. Beale, who, in 1848, represented that district of which Mason County, Virginia, was a part, in the lower branch of Congress, in which he says "I have introduced a bill in Congress asking for $50,000 with which to erect a monument to commemorate the Battle of Point Pleasant." As nothing came of it, it died in a committee room.

That amount of money in purchasing power, equal in value to twice that amount of money at the present time, only demonstrated the magnitude in which the battle was held when not so many years had intervened since that terrible struggle.

We find by reference to an old minute book preserved by Mrs. John Daniel McCulloch, that a monument committee had been organized in the year 1860. The exact date of organization is not given. The first meeting recorded is as follows:

"Monument Association Rooms

Sep. 17, 1860.

The Regent being absent, Mrs. James Hutchinson, Vice Regent, called the association to order.

On motion the minutes of the last meeting was suspended."

"The committee on By Laws reported series of laws by Mrs. Wm. Smith, Chair lady, & under discussion said By laws were adopted,& on motion the committee was discharged."

"On mot on a permanent committee composed of Mrs. Charles Lewis, Mrs. Barlow, Miss Sallie Henderson, Miss Kimberling, Miss Till Stribling, Miss Sue Waggoner to collect historical facts connected with the battle of Point Pleasant.

"On motion of Mrs. Smith, it is resolved this association will celebrate the anniversary of the battle, 10th of October.

"On motion it is resolved committee be appointed to see what will be contributed for a supper.

"On motion a committee composed of Sallie Lewis, Fannie, for the Flats, Miss Patrick, Sehon for Mason City, Miss Stribling & Hall, upper part of town, Miss Jones & Miss Murdock lower end, Sallie Henderson and E. Smith, South Side Kanawha, Ginnie Neale & Maria Menager, Mercer Bottom.

"On motion it is resolved a committee of two be appointed to wait on Col. Beale, & see if we can procure the Hall.

Signed, E. Smith

Recording Secretary

M. T. Lewis Regent

Nov. 14-1860"

There is left no written record of that supper given at Beal's Hall, but there are many living here yet who recall it as one of the greatest social events of the town up until that time, as there was gathered here all of the elite of the county. The money raised at that time by the supper was about $200.00, which was supplemented by $800.00 more in subscriptions, Mrs. John S. Lewis (Mrs. Mary T. Lewis) the Regent riding horseback over the county soliciting funds.

The society applied for a charter which was granted under the laws of the State of Virginia. The money was loaned to Mr. Peter Steenbergen Lewis, a descendant of Col. Charles Lewis killed in the battle, and was faithfully accounted for and interest paid until turned over to the Point Pleasant Battle Monument commission provided for in 1901, by the State of West Virginia.

The only two surviving charter members of the original monument association are Miss Elizabeth Smith, of McCausland, and Mrs. J. D. McCulloch (Miss Sallie Lewis) of Point Pleasant.

We find in the above mentioned record book the following entry:

"Monument Association Room.

Nov. 14, 1860.

The Regent, having called the meeting to order on motion of E. Smith, the historical committee is requested to wait upon, or otherwise communicate with all the early settlers of the county, that is practicable, to obtain all the information they can in regard to the battle of the Point, and all other interesting events of the early Indian times.

On motion it is resolved, the monument be placed on the spot where the brave men who fell in the battle have so long lain unhonored, by vote was unanimously carried affirmative Nannie Smith, Kate Murdock, Sallie Lewis, Sallie Henderson, M. J. Stribling, Ginnie Neale, Rose Barlow, Fannie Lewis, Eliza Waggoner, E. Smith Negative.

"On motion it is resolved a fine of five cents be imposed on those who are not present by half after two o'clock, P. M.

"On motion it is resolved that this meeting adjourn to meet the first Wednesday in January."

It is well that the names of these patriotic women have been preserved to history, many of whom were descendants of participants in the battle. Sallie Lewis (Mrs. J. D. McCulloch) descended from Col. Charles Lewis; Sallie Henderson, the late Mrs. Jos. George, of Five Mile, descended from Samuel Henderson; Misses Sue and Eliza Waggoner descended from Gen. Andrew Lewis; Mrs. Charles Lewis was the mother of Mr. P. S. Lewis, a descendant to whom the first funds were entrusted; Mrs. Kimberling was the wife of Elijah Kimberling, for many years clerk of the county court; Fannie Lewis, wife of Judge John W. English, descended from Col. Charles Lewis, as did Miss Lizzie Sehon, of Mason City; Miss Hall was the late Mrs. B. J. Redmond, daughter of Hon. John Hall; Miss Jones is Mrs. J. W. Bryan; Maria Menger became the wife of Rev. George Lyle; Miss Till Stribling became the wife of Mr. Chap. Waggoner of Pleasant Flats: Mrs. Rose Barlow was the wife of a resident physician; Miss Kate (Beale)

Murdock was the second wife of the late Col. C. B. Waggoner, Ginnie Neale now Mrs. Otis Stribling.

The Civil War breaking out, the efforts to erect a monument were put aside for the stirring incidents then agitating the minds of the people and no efforts were again made until the 100th anniversary of the Battle, 1874, when the proper celebration of the battle was taken up by Messrs. John Q. Dickerson, John D. Lewis, C. C. Lewis, Wm. Dickenson, of Charleston, P. S. Lewis, J. P. R. B. Smith, Judge John W. English, of Point Pleasant. These largely financed the celebration assisted by other of the most patriotic citizens of Point Pleasant and an effort was made to gather together as many as possible of the descendants of the Lewis's of that battle. In "fact so little attention was paid by other descendants and so highly had the Lewis's honored the services of their sires that the proposed monument was spoken, of as the "Lewis Monument" and, for years, the writer, who was present at that celebration, scarcely knew there were other heroes participating worthy of being published in the school histories, there being no available books to be read and no one mentioned by word of mouth but the Lewis's.

All honor, however, to this family who honor their heroic dead. It was the most splendid palm they could place upon the brow of their ancestors to teach the world as they have done their descendants to revere the names of Andrew and Charles Lewis.

The first published agitation for the proper observance of the 100th anniversary of the Battle of Point Pleasant, we find in the *Charleston Courier*, reproduced in the *Weekly Register* of March 19, 1874, which we quote in part:

"It was at this place that occurred one of the bloodiest and severest battles in which the whites and Indians have ever engaged. On the morning of the 10th of October, 1774, an army composed almost entirely of Virginians, under the command of General Andrew Lewis, and numbering about eleven hundred men, was attacked by a largely superior force of savages under the command of the famous chieftain Cornstalk. The battle raged furiously the entire day, and ended in the defeat of the Indians, who throughout the battle, are said to have displayed the most determined bravery. The Virginia army sustained in this engagement a loss of seventy-five killed and one hundred and forty wounded."

"Among the slain were Colonels Charles Lewis and John Field and eight subordinate officers, all of whom were of the best families of Virginia."

"The loss of the savages was never ascertained, as it was their custom to bear off and secrete their dead. Some twenty or more bodies, however, were found on the field, which the Indians had been unable to carry away."

"The wounded whites were placed within entrenchments, thrown up at the point of the confluence of the Kanawha and Ohio rivers, and a garrison left there to protect them, the dead were buried immediately outside of the entrenchments, though in a scattered manner. General Lewis then pursued his march northward."

"Finding ourselves at Point Pleasant with considerable leisure and time, we proposed to an old friend and resident of the "Point" to take us to the graves of the heroes of the battle of Point Pleasant. With a willingness to oblige which is a prominent characteristic of that gentleman, he readily assented, and in a few moments we found ourselves close to the junction of the two rivers, standing on tip-toe looking over a high bank on which we were standing completing the bounds. A few indentations or depression were all that indicated that within that small enclosure were buried some fifty or sixty heroes of the times that tried men's souls. The place was strewn with filth and refuse and seemed to be a general depository for the rubbish of the neighborhood. With a feeling of disgust at the cold neglect so plainly manifested by the authorities, not only of the governments of the States of Virginia and West Virginia, but of the county and city wherein the rest these dead, we turned away."

The writer who signs himself "Virginiout," relating an interview with Mr. Andrew Darst, residing upon the extreme point where the rivers meet, who not only exhibited a grind stone, cannon ball, and shovel, taken from an old well that had been within the fort, but he exhibited the site of the old magazine long since gone over the bank, and the site of the cottage wherein Cornstalk was murdered.

Quoting further from article of date above given, Mr. Darst said in that published interview: "About 1832 thar came by here an old man who had been here in Injin times. Some folks were wondering whar Cornstalk had been buried. The old man said he knowed, and if they'd follow him he'd show 'em. So he took 'em out to a ditch just back of that drug store you see there, (2nd and Viand Streets) told them to dig in at a certain place and they'd find Cornstalk about four feet under ground. They dug in there and sure enough they found him. They then took him up and buried him in the Court House yard."

“The spot of land here on the point was once a big Injin graveyard, and if you will take the trouble to look over the bank where it has been washed you'll find bones a plenty. All of ‘em’s across beads and trinkets among the bones.”

Acting on the suggestion we took a look over the bank and discovered many fragments of bones which were lying loosely on the soil or projecting from the face of the bank. After sauntering around a few moments longer we bade "Andy" farewell and walked off to take a look at the town."

"The heroes of Bunker Hill have their monument to commemorate their deeds, but the brave little band sleeping so silently on the bank of the Kanawha, have nothing but an old decayed, worm eaten fence to mark their graves. Point Pleasant and Bunker Hill, were each fought in the same cause, and those acquainted with the history of "Dunmore's war' will not contradict the assertion that the battle of "Point Pleasant," was really the first battle of the Revolution."

The *Weekly Register* of May 17, 1874, editorially comments on the importance of the battle of Point Pleasant and quotes again from the *Charleston Courier*, as follows:

"Is there any event connected with our past history which so closely affects the people of the Valley as the battle of Point Pleasant, where Virginians bared their breast to protect it from invasions? While Eastern Virginia had her Yorktown, West Augusta had already purchased a victory at Point Pleasant. To no event transpiring within the limits of our State has ever attached the importance and grateful recollections as has to the Point Pleasant battle. It is well suggested then that the people all along our Valley take some steps to celebrate the one hundredth anniversary of this event at Point Pleasant in October next. Our neighbors in Mason will readily adopt the suggestion, as well as all other counties that feel an interest in preserving afresh both the memory of the gallant dead and their resting places."

"For many reasons the Mason county people should take the lead in this matter, and we feel confident they will. Let every community then from the Ohio to the Greenbrier, fall into line and adopt some harmonious action to fitly celebrate the day, and to raise suitable funds to remove the disgrace of the neglected graves. There is not a superabundance of time, and we predict a prompt response from Mason. Who will take the initiatory for a grand celebration of this event, which if of all others the one great shrine which every creed, every political faith and every class in the State can pay their homage."

"A correspondent from Mason County to the *Charleston Courier* contributed the following:

"There is considerable talk just as this time about our centennial anniversary, and a good deal of patriotic feeling is being exhibited in that direction. History records that on the morning of the 10th of October, 1774, there was fought at this Point one of the severest and most hotly contested and bloody battles between the whites under General Lewis and the Indians under command of the great warrior, Cornstalk, that ever took place in the early times of this country. It is in fact, claimed that this was the first battle of the Revolution, and for freedom from the British yoke. On the 10th of October, 1874, one hundred years will have elapsed since that memorable battle, in which the troops under Gen. Lewis achieved such a triumph."

On Tuesday, May 26th, the Directors of the Second Annual Mason County Fair decided to hold their Fair on October 6th, 7th, 8th, and 9th, but no mention is made of the observance of the 10th, the anniversary of the Battle. Plans, however, were going forward from Charleston, as the *Charleston Courier*, in its last issue of May 18, 1774, gives the following:

"The idea of the Centennial celebration at Point Pleasant is a very happy one. It is an event in which every true West Virginian should take pride. Our state embraces a large boundary of territory of "West Augusta," whose sons rendered themselves so famous in the days that "tried men's souls," and to whom the great Washington looked, for raliance when all others should fail him."

"Many descendants of the participants in the famous Indian battle at the Point, are now living in this State. In the counties of the Greenbrier Valley as well as in the Kanawha Valley are living those who bear the name and through whose veins run the blood of the Captain of the Virginia forces, Andrew Lewis, as well as those who descended from the brave men that followed him in that remarkable campaign. The result of the battle at the Point saved all the Virginia frontier from the invasion of the Indians. If Cornstalk had been successful who can imagine the fearful desolation that would have been wrought from the Ohio to the Alleghanies."

"Is it "not a little remarkable that while this battle should have become so famous in history, that so little should be known of the particulars of the fight? While history is silent, we have traditions that should be gathered, and the most authentic ones be placed in some shape as to be reliably transmitted to posterity. There are many households of West Virginia, where stories of grandfather's experience in the battle of the "Pint" are related to day, and many of them told by those who have heard the relation from the lips of the veterans himself. What a pleasant task then for some one to collect these traditions and weave his- tory from them."

"Just as the battle of Point Pleasant was the prelude to the war of Independence, so let the celebration at the Point in 1874 be the prelude to the grand affair to come off at Philadelphia in 1876, and let every West Virginian, and every Old Virginian, and every one who sees proper to join us, take part in the jubilee on the 10th of October next."

To further stimulate the interest in the Battle Celebration, the *Register,* on June 25th, copied from *Nile's Register*, of May 3, 1817, an account of the battle and in the issue of August 27, 1874, the *Register* copied De Mass' *History and Indian Wars in West Virginia*, the account in full of the battle of Point Pleasant. The *Register* of October 8, has tailed up until that time to give any program or details of the celebration, but, in speaking of the Fair then in session, says: "The Fair will close on Friday evening with a grand ball at Beale's Hall. On Saturday the Centennial Celebration will come of.

The issue of the *Weekly Register* of October 15, 1874, gives the following detailed account of the Celebration; written Oct. 10th, 1874.

"The Centenniel celebration of the Battle of Point Pleasant is now over. Just one hundred years ago to-night, brave men and true were mourning over the dead, and ministering as best they could to the wounded and dying. Let us as best we can, look back upon the day that has just been closed by the setting sun of the 10th day of October, 1874, and tell our readers what has been done. Many an eye looked out anxiously this morning to see what was to be the prospect for a beautiful day. For one I was sadly disappointed and feared that the Heavens would soon be sending down the rain.

"How anxiously we watched every appearance indicating like a breaking away of the lowering clouds. Soon after an early breakfast, the clouds began to look thin, and then spot after spot of blue sky was seen. Anxiously did the eager gathering crowd, look for the promised published programme. About 9 o'clock A. M., the Register office sent out the first, which were eagerly seized by the hungry crowd; then another and another handful of programmes were distributed, so that before the hour of 10 A.M., all seemed to be posted as to what was to be done. About 10 o'clock the Company of Cadets from the University of West Virginia, at Morgantown, were formed in line by their Captain, H. H. Pierce, in front of the Kline House, near the wharf-boat, on the Ohio, and waited to receive the Knights of Pythias, from Gallipolis, Ohio, accompanied by the Gallipolis Brass Band. Soon the Knights came marching up, splendidly dressed, and a fine looking body of men they were, passing in front of the Cadets, they halted on Main Street. The Cadets, moved in column of fours up to Main Street, then wheeling to the left, were halted opposite the Court House the site of which is supposed to have been on the line of battle as it was formed, just one hundred years ago today. Here let us give the programme as follows:

Centennial Celebration

of the

Battle of Point Pleasant.

Order of Exercises.

Procession to form in front of Court House at 10 A.M. in the following order: Mayor, Orator, and Committee of Arrangements.

State Cadets.

The Clergy.

Relatives.

Music.

Knights of Pythias and other Orders.

Distinguished Guests.

Citizens.

Funeral Procession.

Escort of State Cadets.

Re-interment of the remains of the heroes who fell in this battle, with becoming ceremonies at 3 o'clock p.m.

Under the effective Marshals who had been on duty at the Fair Grounds for the last four days, the column was formed The Cadets were headed by their own drum corps; the Knights of Pythias by the Gallipolis and Point Pleasant Brass Bands combined, whilst the Ravenswood Brass Band marched up the side walk and waited for the formation of the column, after which it took the place assigned it.

Just before 11 A.M. the column was put in motion and marched in the order assigned, to the Fair Ground, where more than a thousand people were found waiting the arrival of the column. So densly crowded was the amphitheater before the head of the column reached its front that it was with the utmost difficulty that the Committee of Arrangements could clear away space enough for the Company of Cadets. A stand for the speakers had been erected just opposite the center of the amphitheatre. This stand was occupied by the following persons:

Col. Lewis Ruffner, Col. C. B. Waggener, Col. Benj. H. Smith, Capts. H. H. Pierce, Commanding the Cadets, Geo. C. Sturgess, Corresponding Secretary of the Historical Society of the Board of Regents University of West Virginia, J. W. Screntz, Treasurer of the same, Dr. Thomas Creigh, of Greenrier County, Dr. S G. Shaw, President of the Centennial Society, F. A. Guthrie, Attorney at Law, and member of the Committee of Arrangements, Jno. E. Timms, Attorney at Law, Secretary of the Mason County Agricultural Society, Col. (Dr.) A. R. Barbee, G. W. E. Mitchell, of Portsmouth, Ohio, John D. Lewis and Jno. Waddell, who was wearing the shot pouch and powder horn carried by his father, Alexander Waddell, in the Battle of Point Pleasant. The Clergy occupying the stand were Father Francis Guthrie, one of the Pioneer Methodist Preachers of the Kanawha Valley, Revs. S. E. Lane, T. H. Rymer, T. H. Lacy, G. C. Wilding, and W. E. Hill. The exercises were introduced with a prayer offered by Father Guthrie, which was full of thanksgiving and praise, then the orator, Col. Ben Smith, was introduced by Mr. Timms. He commenced reading his well written and interesting address at 12 o'clock and 22 minutes, just five minutes after the 13th gun of the salute was fired by the Artillery Company from Gallipolis, Ohio. This occupied about thirty-five minutes. The speaker took his seat amid deafening roars The following resolution was offered by Dr. Barbee:

Resolved, that the thanks of this audience be returned to Col. Smith, for his interesting address, and that, with his permission, it be published in full in the Point Pleasant Register. This resolution was carried without a dissenting voice. After this the following preamble and resolutions were offered by Rev. W. E. Hill:

Whereas, just one hundred years have passed by since the battle of Point Pleasant was fought and won.

2. In this battle there was displayed gallantry of such character, as to merit a lasting monument.

3. It is the opinion of some, whose opinion is worthy of respect, that this battle bore an important relation to the war of the Revolution, by which the original thirteen colonies gained their independence.

4. The battle of Point Pleasant hastened the material prosperity of this and other States, by the sacrifice of noble blood, therefore be it Resolved.

1st. That a committee of three with power to engage others, be appointed to solicit contributions for the purpose of erecting a monument and purchasing the ground round about the spot where the remains of our heroes now repose.

That this committee be empowered to raise a subscription on the ground today; to write to the descendants of the brave men who were engaged or fell on the field of Point Pleasant, asking them to aid in this work by contributions; to ask the Legislatures of Virginia, West Virginia, Kentucky and Ohio, to make appropriations to this work; and to request also the Congress of the United States to make an appropriation to the same end.

2nd. That this monument be erected within the next twelve months, and of West Virginia marble.

3rd. That it be unveiled on the 10th of October, 1875.

4th. That the committee be empowered to arrange for funeral oration and an historical address on the occasion: and to make such other arrangements as may be necessary to gather together the military organizations of the State; the various secret societies of a benevolent character; the legislatures of the State, &c.

Mr. Hill introduced these resolutions by a motion to resolve the vast audience into a Monumental Centennial Organization. After the reading of the resolutions a motion was made to adopt. Pending this, Dr. Creigh arose and asked to be allowed to speak on the question. Permission was granted, and the Dr. perfectly thrilled those within reach of his voice. After the Doctor's eloquent speech, the motion to adopt the resolutions was carried with but one single dissenting voice, whose "no" was followed by cries "of ,"put him out!" "Knock him down," &c.

Mr. Mitchell, of Portsmouth, Ohio, was then introduced and made a short speech, which could not be heard very far off, owing to the noise of the moving crowd and the low pitch of voice, and its effectiveness was marred to some extent. Loud cries were head for (Walker). It was announced that Mr. Walker was at the Point, and had been sent for, but could not get here for some time.

A cry for Sturgess brought that gentlemen to his feet, and he made a very happy address.

The President, Dr. G. S. Shaw, announced as a temporary committee to wait upon the crowd, to solicit contributions for the monument, viz: E. L. Neale, Dr. C. T. B. Moore, and F. A. Guthrie. The latter gentlemen being a member of the Committee of Arrangements, suggested Rev. W. E. Hill, as his substitute. Recess was taken and ample provisions were made to fill the inner man. Many families gathered in groups about the grounds, in picnic fashion, and ate their dinner whilst a roast ox was served up at the public table. During the recess the committee raised by subscription about six hundred dollars. One gentleman subscribed two hundred dollars in cash on two papers, and we do not think he indends to stop at that if more is necessary from him, to erect the monument. There was such a dense crowd and such hurrying to and fro that it was almost impossible to get the attention of the people long enough to get them to subscribe, or we doubt not, more than a thousand dollars could have been raised.

After dinner the crowd was called together to listen to Hon. Henry S. Walker, who delivered a most appropriate and thrillingly eloquent address. We would not attempt to given even a synopsis of it, so carried away were we with its effect upon the people that we could not take a note but stood, feeling with delight its effect upon our self and watching the feelings of others as their animated countenances told that the touch of eloquence was forcing the tell-tale blood to their faces. Frequent outbursts of applause interrupted the speaker, and a hearty vote of thanks was returned for his eloquent address. We must say one thing about Mr. Walker's address; it was written and written too in a few hours, for he was captured as he was aiming to pass the Point. In this case captured property proved to be A No. 1. After the address by Mr. Walker, the procession was reformed in the same order and took up the line of march from the point at the head of Viand Street the column received the Hearse and Casket containing the remains of some of the heroes of Point Pleasant that were killed in the battle. These were exhumed on yesterday and today, under the supervision of the Committee, Dr. Barbee Superintending in person. Here the Cadets with reversed arms and muffled drums, took their position as an escort. The column moved down Main street, to the Kanawha River, and filed to the right, where the bone of our slain heroes were re-interred with military honors. At the grave the beautiful burial service of the Church was read by the Ministers present in the column.

As the procession was moving down Main street, first the Court House Bell, then the bell of the M. E. Church, South, were tolled, whilst the minute gun was fired by the Artillery Company. The drums were muffled and the band played an appropriate piece, thus giving solemnity to the exercises.

Just before the benediction was pronounced, a vote of thanks was returned to the Morgantown Cadets, to the Knights of Pythian, Band and Artillery, of Gallipolis, Ohio, to the Ravenswood Band, and all others who had aided in the Centennial Celebration.

The crowd then dispersed, the Cadets marched to the Kline House, where three rousing, hearty cheers were given them by all present; the Knights took up their line of march toward their homes, at Gallipolis, and the other organizations went also to their homes. The Cadets took the steamer Clara Scott for Charleston.

Thus ended the First Centennial Celebration of the Battle of Point Pleasant. We say in concluding our description of it, from morning till night; each one that had a given part to perform, vied with the other as to which could do it best. All acted well their parts.

There may have been, but we did not see, a single drunken man on the ground.

The Point Pleasant Brass Band has done better than its most sanguine friends had hoped for it, both during the celebration of today, and the three days of the Fair. May it still continue to improve we can stand the hum drum of practice for the sake of such a treat as they have given us in the last few days.

Let us all now go to work earnestly and determinedly to make the "unveiling of the Monument" as decided a success as the Celebration of today but let us make it wider in its extent. Let us make it State and National. Come up, one and all with what you can do with money and influence, and we shall have the pleasure of chronicling at the end of another year, the success of the enterprise inagurated today. There will be some croaking as a matter of course some men grumble even at the wisdom and benevolence of the Infinite God but let us, who are determined to succeed in the enterprise before us, turn a deaf ear to all croaking, and the more they croak, the more determined let us become.

One hundred years ago was a great historical epoch of Point Pleasant. Today has been another, let one year from today be another. Who dares day "nay?" What citizen of Mason County, or West Virginia, or Virginia, or Ohio, or Kentucky, or the United States, could refuse to aid in doing honor to the heroes of Point Pleasant?"

Brief Mention.

"In the midst of the throng of the Centennial celebration, we noticed many of the direct descendants of the warriors of one hundred years ago, and will recall the names of them as far as we can: There were the Easthams, the Somervilles, Jas. Arbuckle, Jr., of Greenbrier, John D. Lewis, of Kanawha, the Lewises, of Mason, Mrs. Agnes Sehon, who had two grandfathers in this great battle, (Col. Charles Lewis and Col. John Stuart,) who is also the first representative of four living generations who were upon the ground and who are liniel decendents of the two pioneers Cols. Lewis and Stuart. The fourth generation was represented by her great grand child, little Grafton Tyler, who is now some three or four years old. Mrs. Sehon was probably the only person on the ground that could say as much John Waddell, of Ohio, was also present and had swung around his shoulder the powder-horn and bullet pouch, carried by his father in the Battle of Point Pleasant on the 10th day of October, 1774. The Hannans, the Clendennins, the Millers, of Mason County, the Clendenios, the Hannans, the Millers of Ohio. There were a number of others present whose names we did not ascertain."

On October 22nd, the *Register* published the speech of Dr. Thos. h of Virginia, delivered on the occasion of the anniversary of the battle, which we quote in part:

"I am glad I am here because I witness a scene, (and I appeal to the best and oldest historians here present for the truth of the remark, that such a scene of moral sublimity, except the signing of the Declaration of Independence, has never been presented to the people of these United States as is presented here today.) And what is that scene? I see around me some two thousand people with the descendants of the heroes of the battle of Point Pleasant, to the third and fourth generation, on this 10th day of October, 1874, under a bright October sun, one hundred years after the battle, assembled on the battle-field, following the lofty instincts of our nature, to gather the bones of their ancestors, place them in a metallic coffin, catch the inspiration of their ever living virtue and valor, and determining to place a high and enduring monument to their memory. Yes sir, to erect a monument high and lofty on the banks between the Great Kanawha and Ohio, to overlook these two great rivers, the music of whose waters will mingle with the names forever, where the passengers on board your vast merchant steamers, sailing on these great commercial highways, shall see it and as they pass by uncover their heads and toll from the bell a requiem to their memory, and where the valiant youth of our country shall in all future time come to worship at this shrine, study their character, imitate their virtues and be inspired by their lofty deeds of patriotism, and where boys and girls in happy bands shall come for long centuries and cover this marble column with the cypress and the vine, and the immortal flower."

"Yes, sir, build this marble column to their memory, for they were "tall and grand" old warriors. Do you know why I say "tall and grand?" they are not original terms with me. I will tell you where I got them. Old Pool, a colored man, was the body servant of Col. John Stuart, of Greenbrier at the battle and witnessed the fight at Pt. Pleasant, and took part in it. Old Dick Pointer, another colored man, was the hero of Donnalley's Fort, near Lewisburg, Greenbrier county, and for his bravery, was freed by the Legislature of Virginia, and received a pension. Col. Ben Smith, our orator, was no doubt a member of the Virginia Legislature at that time. Old Pool and Old Dick were talking over their battles in Lewisburg one day when I was a school boy, and we school boys were all standing by and listening. Pool and Dick became very much excited in telling what each had done. Dick says the Indian is taller than the white man for I killed Indians as high as the court house. No, says Pool, if you could have seen Mas. John and Charles and James, in the light at the Point, when fire flew out of their eyes like the fire out of their guns; they were just as tall and grand as the old trees on Kanawha. That is the reason why I say they were "tall and grand old warriors."

"But, sir, history informs us that there was a man by the name of George Washington, who had a high opinion of the companies of these Point Pleasant, warriors, for he said in a dark hour of his campaign to his wearied and dispirited army on the plains of New Jersey, only give me the men to place the standard of my country on the mountains of West Augusta, and I will call around me the men who shall make my country free. These 'tall and grand' warriors were West Augusta men, and Washington knew them."

"Yes, build a monument of enduring marble to the memory of those old grand warriors, for you may look down the long picture gallery of history and you will find no brighter or grander names than the men of 1774 and 1776. Nor let us forget their characters, or their great principles of civil liberty, or insult the spirit of liberty."

While the *Register* of December 10, 1774, gives the following account of the committee.

The Point Pleasant Monument Association held their first meeting in the law office of John W. English, Esq., December 4th, 1874. All the members of the Committee, viz: W. E. Hill, F. A. Guthrie and John W. English, were present. On motion, John W. English, was called to the chair; and on further motion, was made permanent Chairman. On motion, C. E. Hogg, was elected Secretary, and T. Stribling, Treasurer. On motion, the Committee was ordered to inquire into the title of the land on which the monument is to be erected. On motion, the Committee was directed to see Drs. Shaw and Moore with reference to subscription papers, and also ascertain how much money is in their hands belonging to the Association, and to pay the same to the Treasurer. On motion, the Association adjourned to meet Thursday evening, December 17th, 1874."

While no report has been preserved of the Monument Association, founded on October 10, 1874, the work of procuring funds was taken up by the Mason County representatives in the Legislature of West Virginia, Hon. Edmund Sehon, descendant of Col. Charles Lewis introducing the bill in the lower house, while Hon. P. C. Eastham, descendant of George Eastham, of the Battle of Point Pleasant, introduced the bill in the upper house. On Feb. 25, 1875, the Legislature passed a bill carrying an appropriation of $3,500.00 to aid in the purchase of land and the erection of a monument in commemoration of the Battle of Point Pleasant, the President and Secretary of the Monument Association to have charge of the erection of said monument, the parties investing the money with approved security, awaiting assistance from other states. No further action was taken until Feb. 26, 1897, when the Legislature adopted a Joint Resolution by which Governor Geo. W. Atkinson appointed Judge John W. English, Dr. A. R. Barbee and Judge F. A. Guthrie as custodians of the fund appropriated by the Legislature.

The matter again lay dormant and no effort was made either to collect the funds or secure further appropriations.

That the State and Point Pleasant was not free from censure for their dereliction of duty, we quote in part from the *Kingwood Argus* of June 29, 1899, copied in the *State Gazette* July 4th, of that year: "We were at Point Pleasant and visited the graves dead who fell in that memorable battle and we were surprised and indignant to find the place almost surrounded by stables and hog-pens and lying along a back alley with not a mark of any kind to designate the spot; not even a fence or wall around the place and no one could find it without a guide who knew just where it was. It was only a very small plot of ground, in some one else's back yard, with stables on two sides of it and a garden on another side and the only way to get to it is down a dirty alley. Nearly a hundred pioneers of that section who fell in that battle, fighting old Cornstalk for possession of the beautiful land along the Ohio river lie buried there, unmarked and almost forgotton. It is a burning shame and disgrace on the town to allow it. We went up to a newspaper office and made a vigorous kick about it and learned that money had been appropriated by the State and also quite a sum raised by the ladies of Point Pleasant to erect a monument to these heroes, dead defenders of our country. Now let justice and decency compel a disposition of this matter. The State or the town should take charge and secure some contiguous ground to the resting place which cannot be called a cemetery or even a graveyard, as it is now, and make a little park out of it with a fence around it, and erect a handsome monument on the center of the site and make a decent way to get to it by removing some of the adjoining stables and hog pens."

The *State Gazette* appended the following editorial comment: "We think the suggestion that the park and monument should be contiguous to the resting place of the dead heroes and should include it as well as the site of the old fort is the correct solution. The *Argu*s will no doubt be surprised to learn that there are suggestions now made that the monument be placed away up on the Ohio bank twelve squares from the site of the old fort and grave of Col. Lewis and others, and suggestions that it be placed back on the hill overlooking the town. "

There is no mention of the celebration of the anniversary of that year 1899 save the local newspaper account that,

"Today, October 10, 1899, is the one hundred and twenty-fifth anniversary of the battle of Point Pleasant, fought between General Andrew Lewis and Cornstalk, the sachem of many nations and no citizen evidenced any interest save Col. G. B. Thomas, the Post-master, who decorated the Post Office with flags."

No further agitation of the subject appeared in the local newspapers save that of the *State Gazette* of April 21, 1899, urging the organization of a Chapter of Daughters of the American Revolution and Sons of the Revolution here, looking forward to proper recognition of the Battle of Point Pleasant in which it said: "Here was fought the first battle of the Revolution and why should not this town and county boast of the largest patriotic organizations in the State."

No further steps were taken at Point Pleasant looking to a monument appropriation until Feb. 11, 1901, when Mrs. Livia Simpson-Poffenbarger issued a call for the organization of a Chapter of the Daughters of the American Revolution which is of date of Feb. 14, 1909, reported as follows in the *State Gazette*:

"A sufficient number of ladies met at the home of Mrs. George ' Poffenbarger yesterday to organise a Chapter of the Daughters of the American Revolution. As soon as the requirements are duly met, a Charter will be granted. This is a most commendable move and should meet with the support and best wishes of the entire town."

At this meeting it was explained that the chief motive in organizing this patriotic society was that an organized effort might be made to secure funds for the erection of a Battle Monument at Point Pleasant and the recognition of the battle as that of the first Battle of the Revolution.

On Feb. 26, 1901, a call was issued for the organization of a Chapter of Sons of the Revolution and the *State Gazette* of that issue said editorially, "This is a move in the right direction. This is the historic spot of the first battle of the Revolution. In order that history may accord us that place with due credit, we must first show that Point Pleasant appreciates this fact. If the spot is to be properly marked and receive from the National Government the appropriation and recognition for a splendid monument that the battle deserves, we must first show that the memory is dear to us."

On Feb. 29, 1901, the Daughters of the American Revolution were formally organized and the name selected was that of Col. Charles Lewis, named for one of the best beloved heroes of those who fell at the Battle of Point Pleasant. At that meeting the Daughters in conformity with the statement of Governor White that he would appoint one man outside of Point Pleasant and two resident members of the commission to look after the funds the ladies recommended V. A. Lewis of Mason and P. S Lewis and C. C. Bowyer of Point Pleasant, but the Governor later decided to appoint but one man in Point Pleasant and the commission named was John P. Austin, C. C. Bowyer and V. A. Lewis. On March 11, 1901, Mr. Joe Friedman not only presented the Col. Charles Lewis Chapter the money with which to pay for their charter, but also for the purchase of their record books and proffered when a monument was completed to donate a splendid Band and Speaker's stand.

The *State Gazette* of March 11, 1901, says: "The new committee to be appointed should first secure the grounds. This would be the beginning looking toward the end. Then the old buildings could be cleared away, grading done, grass sown, trees set out and the grounds beautified. These all take time to bring them to perfection. There is sufficient money all ready subscribed, together with what could be secured, by private subscription to do this much handsomely. The government could then be presented the ground as a National Historic Park and the War Department under the head of Rivers and Harbors could be induced to grade the banks bordering on both rivers and they would then be forever taken care of. Besides, its historic significance, it is in point of beauty of location the most desirable site in town. The view is splendid from both rivers. The view from the surrounding hills is perfect and it would give strangers a different opinion of the town from the ragged appearance it now has from rail and river.

On April 18, 1901, the newly appointed Monument Commissioners gave bond before the Mason County Court with approved security and when they had met for organization, John P. Austin was elected President and C. C. Bowyer Treasurer and V. A. Lewis Secretary.

The issue of May 28, 1901, of the *State Gazette* says: "The Battle Monument Commission held a meeting at the Merchants National Bank Saturday last. The members composing the commission Hon. John P. Austin, President, Hon. V. A. Lewis, Secretary and Hon. C. C. Bowyer, Treasurer, together with other citizens and a *State Gazette* Reporter went down to the Kanawha point where had stood the old fort and where some of the heroes of the battle of Point Pleasant were buried and made a careful inspection of the grounds. The public seem generally united in the belief that this is the proper place to erect the monument and it might be said that the property bounded by Main and First Streets and the two rivers will be purchased shortly by the commission provided the owners of the property do not demand too exorbatant a price."

On Decoration Day, May 30th, 1901, for the first time a public memorial exercise was held, whereby the graves of the heroes of the battle of Point Pleasant were decorated, in charge of the D.A.R. and G.A.R. societies of the town. A large concourse of people attended the exercises. The *State Gazette* of June 18, 1901, records that John D. McCulloch had given an option on his property on the monument site for $1,000 00; C. H. Varian $1,200.00; Geo. T., Chas. and Henry Stone $3,000.00; Thos. Durst $600.00, a total of $5,800.00, which includes all the contemplated territory needed except the Geo. Comstock property, upon which a price had not been agreed."

Later Mr. Comstock's property was secured at $2,200.00.

From the issue of August, 1901, of the *Charleston Daily Mail*'s report of the Monument Commission, held at that place we glean the following: "Hon. V. A. Lewis reported that the Commission had about $11,000.00 in the treasury, but that the work, as planned would necessitate the expenditure of $25,000.00 more." Speaking of the Battle of Point Pleasant, the *Daily Mail* quotes Mr. Lewis as follows: "All careful painstaking thoughtful historians have regarded it as the first in the series of the Revolution which gave the continent to liberty. It was the chief event of Dunmore's War."

On August 29, 1901, the *State Gazette* announced that the Committee had decided to commence clearing the grounds at once and that paper made the first appeal for the Celebration of the Battle.

"It is the intention that the clearing of the grounds shall be completed by the anniversary of the great battle of Point Pleasant, Oct. 10, 1774."

"The significance of the battle has not been wholly overlooked. Thousands know of its importance and it remains for Point Pleasant herself to appreciate her relation to history and demand from the State and from Congress the substantial recognition due this spot. Our citizens should join in one glorious celebration of this anniversary Oct. 10. We trust that there is enough patriotism in the town to observe it. If in no other way, let it be one grand union picnic. The weather will be fine and all can come together at least in the spirit of patriotism and good fellowship".

Mrs. Poffenbarger, editor of the *State Gazette*, not only issued a call for a citizen's meeting on Thursday night September 5th, looking toward the celebration, but she had secured the cooperation of Col. J. P. R B. Smith and, at that meeting donated the services of her paper to advertise the meeting, supplemented by a subscription secured by her of over S200.00 with which to begin the work. The paper of that issue contained the following full page advertisement, besides the names of the donors of cash:

127th ANNIVERSARY

The First Battle of the Revolution to be celebrated at

POINT PLEASANT, W. VA.

Thursday Oct. 10th, 1901.

Great National Speakers will be present.

Entertainment for the people. Excursions on all railroads and steamboats will be arranged for. The Old Log Mansion built in 1796, that has lived in three centuries will be used to exhibit the greatest lot of Historic Relics ever brought together in West Virginia, outside the Historical Society at Charleston.

EVERY ONE INVITED.

The monument Park and Court House yard will make fine picnic grounds. Watch this space for attractions as they are secured."

As a result of the meeting at the court house committees were organized and Mrs. Poffenbarger participated in the work of all the committees. The issue of September 19 shows an additional subscription, making a total of $409.00, besides generous subscriptions, of exhibits for the museum. The children of the public schools contributed S8.50 to be used in decorating a wagon for the parade. The newspapers of the country stood up and took notice of the big celebration and helped advertise it.

In the issue of Oct. 10th, The *State Gazette* announced the presence of distinguished visitors, among whom was Gen. C. H. Grosvener of Ohio, faithful advocate for an appropriation from Congress to commemorate the battle. Virgil A. Lewis, in an article published in the *State Gazette* of that date, said:

"After all, even though it be here, is it best to assert without reference to the proof that the battle of Point Pleasant is the first battle of the Revolution and then array against us the whole of New England where the people are jealous of the claims of Lexington. It is easy to make assertions, but to examine hundreds of volumes and obtain records from both Europe and America in proof of the same, is quite another thing. Do not throw the burden of proof of this matter on a committee before a Congressional Committee."

The *State Gazette* of October 17, 1901, announced that the Mansion House had been turned over to of three ladies who had accepted it as a headquarters for the Col. Charles Lewis Chapter D.A.R. Also that work on Tu-Endie-Wei Park had been discontinued for want of funds and made an appeal that the Commission set out trees on the edge of the Park or permit the citizens to do so. Also the list of subscribers to the expense of the Celebration.

While the following is the published account of the big celebration:

"THEY CAME.

From Every Direction.

Ten Thousand People Celebrated the 127th Anniversary First Battle of The Revolution. Tu-Endie-Wei Park.

On last Thursday the good citizens of Point Pleasant celebrated the 127th anniversary of the battle of Point Pleasant, the first battle of the Revolution.

It only needed for our people to arouse themselves and make the effort to have one of the biggest celebrations ever held in the state. Beside our people, who were for the most part upon our streets, enough more swelled the crowd until we had fully ten thousand celebrating. Heretofore, when we had any demonstration in the town, the crowd was from the Ohio river landing to the Court House. This time the streets were crowded from the Court House to the Kanawha River, with fully five thousand people upon the Park.

The K&M Railroad ran a dollar excursion from Athens to this place, and Point Pleasant can never forget the kindness of that road. When we solicited them for a cheap rate they replied if it would be any accommodation to Point Pleasant and they only made it pay expense of running the train they would give it to us to show to Point Pleasant the kindly feeling of that road toward our town. They not only gave us just the service we asked for but they put out 15,000 attractive hand bills and advertised their rates in the newspapers along the route, and when they came in here they brought us not only the Nelsonville and Middleport bands, but they brought a big train packed with people from Athens and way stations; they brought us Gen. Chas. H. Grosvenor, one of the stalwarts sons of the Revolution who came to address our people and promote the success of the day. The K&M brought us a train of eight or ten coaches from Charleston that had standing room only and while the excursion was profitable to the K&M and we are glad it was, we must not forget that it was run wholly to compliment Point Pleasant. The conduct of the K& was greatly in contrast to that of the Ohio River Division of the B&O R.R, who were importuned by letter, in person, and by telegraph, to give us a rate, but of no avail, and notwithstanding full fair was charged, that road brought in crowds of people with a crowded train from the north and standing room only coming in from the south end. Steamboats brought in excursions and hacks run between here and Gallipolis while six hundred persons crossed the Kanawha Ferry, and as many more had to be ferried in private boats in harbor. The farmers came in carriages, buggies, expresses, big wagons, horseback and on foot, any way so they came, and they came as a multitude.

“Notwithstanding the rain, the night before, which was just enough to lay the dust, the sun shone out about eight o’clock and the day was ideal.”

“Never before did the old town wear such a gala dress. Old Glory and buntings galore waved from every residence and business house. There was neither the difference of politics or religion or even the distinction of secret organizations to mar the occasion, but all came together upon a common level for one grand glorification of the celebration of the battle, the farthest reaching in its effect of any battle ever fought upon the American Continent – the first battle of the Revolution – the battle that broke the power of the red men in America; the battle that brought the treaty that enabled civilization to march on to the west and southwest and great northwest territory; the battle that resulted in ceding to Virginia and thence to the colonies the great Northwest; that battle that defied at its close Tory misrule, the first battle ever fought after the tea had been thrown overboard at Boston Harbor the preceding March. The Boston Port Bill, of May, 1774, the signal for actual conflict had been passed. The House of Burgesses, of Virginia, had declared the first of June of that year to be "A day of fasting, imploring the Divine interposition to avert the heavy calamity which threatened destruction to their civil rights and the evils of a civil war." Massachusetts had passed resolutions deploring the oppression of Great Britain. Patrick Henry had made his famous speech before the House of Burgesses, of Virginia, declaring that "The war is inevitable, and let it come," and asked "Is life so dear or peace so sweet as to be purchased at the price of chains and slavery?"

"England, too, recognized before the Battle of Point Pleasant that the war was inevitable, and determined to keep the colonists so busy defending themselves from hostile Indians armed with English muskets and English amunition, that they would have no time to think of the wrongs inflicted upon them by the mother country. Great Britain never had a better tool than Lord Dunmore, the Tory Governor of Virginia, as his subsequent conduct proved. Hence, the battle of Point Pleasant, (in which Lord Dunmore intended the flower of the Colonial Army of Virginia to be destroyed but which, victorious to his surprise) became the first battle in which the blood of patriots was spilled upon American soil for the cause of National Independence, and was so credited by Alexander S. Withers, in his chronicles of Border warfare, later by Bancroft, the Government historian, by President Roosevelt, in his *Winning the West*, and by many other historians, of repute.

Is it any wonder that we point with pride to this battle and invited the whole country to celebrate with us?

"At ten o'clock the parade began to form on First Street and it was of such magnitude that it took Col. J. P. R. B. Smith, the Grand Marshall of the day, and his corps of splendid assistants, an hour to get them all in line ready for march. It was headed by James Somerville, of Pleasant Flats, dressed as Uncle Sam, and in face and figure he is typical of the figurative head of the nation,) all the city fathers, except Mayor Somerville, who acted as a marshall, were in the parade in carriages, The fire department were out with their wagon decorated. There were floats galore, put in by the business men of the town, vying with each other to see which could make the finest display. The three splendid bands, Cheshire, Middleport and Nelsonville, discoursed sweet music along the route and the children of the white schools marching in line wearing bouteniers of national colors, waving flags, which little girls representing "the states and territories, an d the colored children on a wagon beautifully decorated made the prettiest parade ever seen in Point Pleasant.

What might have been a serious accident, but proved to be a fortunate escape, occurred when a wagon with 103 children on it passed over a culvert on 14th Street. The culvert went down and the top of the wagon was separated in the lunge from the platform and the children were precipated to the ground. Fortunately no one was hurt and the procession proceeded down Main Street for First, the children falling in line with the others in the line of march, where they disbanded. Three open air concerts were given from 1 to 2 o'clock, when the speaking began from a platform at the grounds recently purchased as a site for a monument yet to be arected [sic]. The grounds are situated at the junction of the Kanawha and Ohio Rivers, without doubt the most beautiful site for a park on the Ohio river."

The large audience was called to order by Col. J. P. R. B. Smith, who called our distinguished fellow townsmen, Hon. C. E. Hogg to the chair. Mr. Hogg in his usual pleasing manner did the honors of the occasion with credit to himself and to Point Pleasant. Mr. Hogg then introduced Mrs. Livia Simpson-Poffenbarger, Regent of Col. Chas. Lewis Chapter of the Daughters of the American Revolution, who had been designated by her Chapter to dedicate and name the park which she did in a short address, as follows:

"Mr. Chairman, Ladies and Gentleman:

It has been deemed fitting and appropriate, that, by some means this beautiful and historic spot of ground be dedicated to the noble purpose for which it has been purchased, and given a name by which it may be known in the future. The important duty has not been sought by the organization I have the honor to represent. I wish to emphasize the fact, without going into explanation or detail, that is has been rather thrust upon us. We are simply doing that which has been denied to others who have been invited to do it, by their situation and present circumstances. We have accepted the trust and assumed the duty in the absence of others who might, and we sincerely believe, would have performed it better.

However, I wish to premise that it is not at all inappropriate that the Society of the Daughters of the American Revolution perform this most important function. Ours is purely a patriotic organization and our work is carried on in the name of patriotism and inspired by love of country. The objects and purposes of our society are set forth in our constitution, Article 1, as follows:

(1) "To perpetuate the memory of the spirit of the men and women who achieved American Independence, by the acquisition and protection of historical spots, and the erection of monuments; by the encouragement to historical research in relation to the Revolution and the publication of its results; by the preservation of documents and relics and of the records of individual services of Revolutionary soldiers and patriots, and by the promotion of celebrations of all patriotic anniversaries.

(2) To carry out the injunction of Washington in his farewell address to the American people, "To promote as an object of primary importance, institutions for the general diffusion of knowledge," thus developing an enlightened opinion, and affording to young and old such advantages as shall develop in them the largest capacity for performing the duties of American citizens.

(3) To cherish, maintain, and to extend the institutions of American freedom, to foster true patriotism and love of country and to aid in securing for mankind all the blessings of liberty."

"Another thing I wish to impress upon all here today is the fact that ours is the only society professing to be founded exclusively upon our Revolutionary struggle that recognizes the Battle of Point Pleasant as a part of the war for American independence. Reputable historians, including Bancroft, President Roosevelt and others have asserted that it was the initial, the first battle of the Revolutionary war. Moreover, they have produced the indisputable evidence upon which the assertion is based. What the concensus of American opinion will be as the years shall roll on and historical research shall bring to light the whole truth, we cannot say. If the verdict shall be the affirmative of that proposition then the first battle shall not be lacking in display of heroism and patriotism, exhibited in the midst of an almost interminable wilderness and hand to hand with a savage and at the same time valorous foe."

"The memory of that great struggle, will we think, be well and fittingly preserved upon these grounds. A splendid and enduring monument is to be erected commemorative of the battle. On some part of it will be a bronze statute of the heroic Andrew Lewis, the commanding general. On it will be inscribed in imperishable letters the names of the brave Col. Chas. Lewis and Col. Fields and all those who fell with them in defense of liberty and the homes of our race. On these grounds will be laid down and preserved the outlines of old Fort Randolph.

Without some reference to the stubborn foe which drew the brilliant flash of fire from the steel of these heroes, in the shades of primeval forests, far from the abode of any white man, this history written in grounds, stone, marble and bronze would be incomplete. The red men were fighting for their homes and hunting grounds. From their stand- point, their conduct was patriotic. They were defending the graves of their fathers."

'To the end, therefore, that history, as far as possible, may be fully preserved and patriotism, in its broadest sense may be recognized, it has been decided to give this park the oldest first name it has ever been known to possess its Indian name. By authority of the Monument Commission appointed by the Governor of this state, and in the name of the Society of the Daughters of the American Revolution, we now dedicate this park, the property of the State of West Virginia, to patriotism and the preservation of history and name it "Tu-Endie-Wei Park;" which signifies in the Shawnee tongue "the mingling of waters," this being the junction of two rives."

Mr. Hogg next introduced Col. Bennett H. Young, of Louisville, Ky., who had been previously invited to address the Assembly, and no happier selection could have been made. The people as a unit fell in love with the man. He talked directly to their hearts, and we fortunately secured his speech which is reproduced in this issue of the *State Gazette*. Col. Young also loaned, for the occasion, his pioneer suit, which consisted of a hunting shirt and flint lock gun which bad belonged to Daniel Boone, which he had completed by the addition of a "long knife," leather breetches and a coon skin cap. Herman Snyder was selected to wear the suit, being smooth shaved and corresponding in weight and height to Daniel Boone."

"Col. Young was followed by our poet laureate, Louis Reed Campbell, who recited in splendid style his poem, written for the occasion:

OUR HEROES.

Grave by grave, where the rivers meet, and gently flow,
The patriot sleeps, and by his side a vanquished silent foe.
Year on year with wondrous swiftness glided by,
And yet no stone was reared where brave men dared to die.

Time's hand was in the game that drove the Indian from his land,
In it the shadow is a wrong that greed could not withstand,
More than a century gone ere right doth o'er wrong prevail,
Alike we honor, now, who face the feathered shaft and rifles' deadly bail.

Departed chieftan of a mighty race, so soon to disappear!
What does the future hold, save memory, softened by a tear?
For even now adown the changing slope of fleeting time
The painted warrior glides away, to leave no trail behind.

No power can dim the luster now of that victorious band,
Who fought and fell and fired again where now we stand.
If fairness to the foe is due, what honor must be theirs,

Whose names too sacred for an eulogy, drift upward with our prayers."

The following is taken from the *Point Pleasant Observer* of October 17, 1901:

"TU-ENDIE-WEI PARK."

A Great Outpouring of the People at its Dedication Last Thursday. A Gala Day For Point Pleasant.

"We have not space to tell of the big celebration of Thursday the 10th. This means that if we were to take up every inch of space in the whole paper we could not tell all about it. So we will have to tell a little about it and let the rest go."

"Day dawned bright and clear with never a cloud to remind one of the little sprinkle of the night before, the town was profusely decorated with flags and bunting. The first signal of the coming dawn was announced by the watchman on the site of Old fort Randolph, who was in charge of the relics, firing the morning gun. Immediately after this the church bells begun to ring, say at 5 o'clock, and the mill whistles and those of the steamboats began to blow until not only everybody in town but every one within five miles of the town, was wide awake.

By eight o'clock, the country people were pouring into town in streams from every direction. All roads led to Point Pleasant and all roads were full. At half past nine the trains began to pour their loads into town and by 10 o'clock the crowd was variously estimated from 6,000 to 10,000 people. At 10 o'clock Col. J. P. R. B. Smith began to form the huge mass of people into a line of march and even with his able assistants it was an Herculean task. Finally the parade was formed and headed by Col. James Somerville and a platoon of police followed by Grand Marshall J. P. R. B. Smith, the large procession moved over the principal streets of the town and at last ended at the beautiful park at the junction of the two rivers."

"The parade consisted of the town officers in carriages, citizens in carriages, citizens on horseback, people on foot, the children of the public schools, both white and colored, some in wagons and some on foot, with three brass bands, interspersed at regular intervals, and all interspersed with display made by our merchants until the whole presented a scene of beauty. Merchants from other towns who inform the newspaper people of Point Pleasant that it does no good to advertise here were also glad to have a place in the big parade and were not ashamed to be the lustiest howlers for Point Pleasant's big day."

After the people had gotten something to eat they again assembled at the park and were called to order by J. P. R. B. Smith who called Hon. C. E. Hogg one of this district's distinguished ex-congressmen to the chair. Mr. Hogg after a short but eloquent address introduced Mrs. George Poffenbarger, Regent of Col. Chas. Lewis chapter of the Daughters of the American Revolution who, in a short address, and by authority of the Monument Commissions named the beautiful plot of ground "Tue-Endie-Wei" Park. "Tue-Endie-Wei" in the Shawnee language means "the mingling of the waters."

Chairman Hogg next introduced Col. Bennet H. Young of Louisville, Ky., after which Louis Reed Campbell recited an original poem written for the occasion entitled "Our Heroes." Next followed Gen. Chas. H. Grosvener, of Ohio, who delivered a characteristic Grosvener address. The program of the day was concluded by the burial of the remains of "MAD ANN BAILEY," which had been disinterred from their resting place of 76 years and brought here for burial, thus carrying out her desire, expressed more than three quarters of a century ago, to be buried on Virginia soil."

"The museum contained large number of historic relics which held the interest of the vast crowd from early morn to the leaving time of the late train and steamboat in the evening."

The *Huntington Advertiser* of October 11, 1901, said,

"The Huntingtonians who visited Point Pleasant yesterday returned last evening happy over the exercises of the day."

"It was the proudest and most memorable in the life of that historic community. Amid an assembly of ten thousand persons, many of whom had come half way across the continent to be present, the beautiful grove at the junction of the Kanawha and Ohio rivers, was for ever dedicated to the memory of the gallant Virginians who, one hundred and twenty-seven years ago, gave to the world the first manifestation of that valor, which in after years, broke the power of Great Britain and made this continent the abiding place of civil liberty."

"No spot in the Ohio valley is so full of historical significance as the old town of Point Pleasant. No spot will be more beautiful when the towering granite shaft, to be erected at an estimated cost of forty thousand dollars, shall greet the eye of the stranger as he approaches this true cradle of American liberty."

Quoting from the *Gallia Times* of October 9th, we find,

"Much interest is being taken by the Point Pleasant people in the coming anniversary of the Indian battle fought there. This was on October 10, 1774, or 127 years ago, and, on this day the power of the Red Men in the Kanawha and Ohio Valleys was wrested from them. The day should long be remembered and we believe will be fittingly celebrated by our neighbor city."

The *Gallipolis Journal* of October 9th, said,

"BATTLE

Of Point Pleasant will be

Celebrated on

Thursday next."

"Our neighbors across the river are making extensive preparations for the celebration of the battle of Point Pleasant, the first conflict of the Revolution on Thursday Oct. 10. It was in 1774 that the most desperate Indian battle on record was fought between the confederated Indian tribes, under the celebrated chief Cornstalk, and the Virginia Militia under command of Gen. Andrew Lewis."

"There will be a relic display and a big parade. The old log mansion that has stood in three centuries will be among the other relics of primitive days. Gen. Grosvener and Col. Bennett Young, of Louisville, one of Kentucky's most magnetic speakers. will be present at the celebration. The remains of Ann Bailey, the heroine of the revolution, have been disinterred and will be consigned to their new resting place on Thursday. There will be three bands and a most interesting program and the 10th promises to be an eventful day at Point Pleasant."

The following is from the *Weekly Register*, of Point Pleasant, W. Va., under date of October 17, 1901.

"THE CELEBRATION.

Of the One Hundred and Twenty-Seventh Anniversary of the Battle of Point Pleasant. Thousands of Visitors Within our Gates. The Celebration a Success in Every Particular and the day will be long remembered by all present."

"Despite the cloudy weather and rain of Wednesday, Thursday morning, October 10th, 1901, dawned with the old Sol, and a twinkle in his eye, to the gratification of the committees in charge and the citizens of our little city in general, for clear and favorable weather meant for the celebration of the first battle of the Revolution, (fought at Point Pleasant on October 10, 1774, between the whites under command of Colonel Charles Lewis, and the Indians,) unbounded and unprecedented success.

The committees in charge of this, the 127th anniversary of this great battle, have been earnest workers to accomplish this end, and deserve much credit for their untiring efforts.

The genial Marshall in chief, Colonel J. P. R. B. Smith and his assistants, handled the big parade which was formed in the first ward in a manner creditable to a general with an army of 100,000.

In the parade was a representation of all our business people, the children of our public schools turning out in full, dressed in national colors, companies of horsemen, the fire department, the hook and ladder company, and everything and every body to make up a grand trades display the like of which has never before been seen in Point Pleasant."

"To say the town was galy decorated with bunting, flags, &c., is not necessary, for the emblem of our great nation was never so profusely displayed as on this occasion."

"The park at the confluence of the Ohio and Kanawha rivers where the monument to commemorate this great battle will be erected, has been put in the best proper shape, and was thronged with visitors and sight- seers throughout the day."

"The old house, considerably over a hundred years old, which was converted into a museum and filled with relics, was the most interesting feature of the occasion."

"The ladies in charge of the museum are commended by our citizens and visitors upon the manner in which this "relic store" was conducted, and the courtesies accorded all."

"A collection of relics, such as Indian implements of warfare, old pictures, dishes, dresses, jewelry and many other articles too numerous to mention, are not seen by one generation more than once, and one who missed this treat is at loss to know or conceive the manner in which these old settlers lived and had their being. Bands of music delighted the visitors and the day was one of joy from morning until night. The crowd was orderly well behaved and jolly. No congregation of human beings had a more joyous time for one day, than did this one."

"After the parade, which moved at noon, and which was never surpassed as a trades display in this place, had disbanded and the throng of people had dined, the speaking at the park began."

"Hon. Chas. E. Hogg, one of West Virginia's most eloquent orators, had charge of the ceremonies at the speaker's stand, and his introductory remarks were well received. Mr. Hogg never fails to please and enlighten his hearer, and at the conclusion of his remarks introduced Mrs. Judge Poffenbarger, who opened the speaking with an address listened to by the throng of people with marked attention, and which was eloquent and instructive."

"Col. Bennett Young, of Jassamine County, Kentucky, member of Congress from his district, was introduced and to say his address was eloquent, logical and interesting, is but a triffle, for the marked attention of his hearers was undisturbed. He is a fluent speaker and one would never tire listening to him."

"Next to be introduced was Mr. Lewis Reed Campbell, who recited a beautiful poem entitled "Our Heroes." This was one of the most beautiful, and being prepared as it was, only a few hours before the hour for delivery, places our young friend at the maximum. The manner in which this beautiful poem was delivered was pleasing in the extreme and delighted the immense crowd who listened with marked attention." "Gen. Chas. H. Grosvener, of Ohio, was next introduced, who made, as all know, an address second to none. The General is one of Ohio's influential and honored sons, and when he appeared on the speaker's stand, a smile of pleasure and gratification covered every face in the audience. General Grosvener's remarks were listened to attentively, and his eulogy to our martyred President, was pathetic. Concluding his remarks, General Grosvener said it would be well for Great Britain, had she a commander like "Cornstalk" at the head of her forces in South Africa, which was met with a round of applause."

"Space and time will not permit us to report this celebration as we would like to, but those who were here had a good time, enjoyed themselves and left our little city with the wish to soon return."

"The short-comings of our report of this big day are attributed to the absence of the editor in chief, and had he been at the helm, our "chases" would not have held the flow from his pen and our "machines" would have been sorely overtaxed."

The *Mail Tribune*, Charleston, W. Va., Oct. 12, 1901, says:

"POINT PLEASANT

Celebrates Important Anniversary. Remarkable Demonstration in Honor of the Famous Battle.

Col. Bennet Young, of Ky., and Gen. Grosvener, of Ohio, Principal Speakers."

"Point Pleasant, W. Va., Oct. 10th, 1901 was the proudest and most memorable in the life of this historic community. Before an assembly of 10,000 persons, many of whom had come half way across the continent to be present, when the beautiful grounds at the juncture of the Kanawha and Ohio rivers was for ever dedicated to the memory of the gallant Virginians, who, 127 years ago gave the first manifestation of that valor which, in after years, broke the power of Great Britain and made this continent the abiding place of civil liberty."

"No spot in the Ohio Valley is so full of historic significance as this old town of Point Pleasant, and no spot will be more beautiful when the towering granite shaft to be erected at an estimate of $50,000.00, shall greet the eye of the stranger as he approaches this true cradle of liberty."

"Gen. Chas. H. Gresvener, of Ohio, in his speech, said in part:

"It is pleasant on this October day to reflect that here on the banks of the beautiful Ohio, then so remote from the center and homes of our ancestors as to be terra incognita to the people of our country and of the world there should have been struck that which turned out to be the first great blow for American Independence and American Liberty. Figure it as you may, the battle whose anniversary we here today celebrate was the first real blow of the Revolutionary War. Nobody so understood it. Grant that; who understood what was to flow from Lexington or Concord or even from Bunker Hill? What was the name of the man who foresaw when the spring time grass of Massachusetts was reddened with the blood of patriots at Lexington that the blood was to sanctify the soil and result in the rights of the people for self government. Lord Dunmore was loyal to the source from which he derived his official dignity and official position. He was how- ever, as appears by the records, not quite the open manly frank man that Americans love to recognize and honor. When he came to the mouth of the Big Hock-hocking river, under promise to join the troops that he supposed were subordinate to him, he failed to join them for the manifest purpose of weakening their forces and thus enable the Indians to overcome the settlers. While he was not acting that manly, open and above board part which Americans love to honor and recognize, but if the reports are true and conclusions are allowable, Lord Dunmore was guilty of an act of the basest treachery and a manifest purpose of the most inhuman outrage. But it cannot be lost sight of that even in this, shameful as his conduct was, he was acting in furtherance of his purposes to aid the government to which he owed allegiance. It may be that in his horoscope he saw the coming of the overthrow of British power in the colonies and the organization of a new government and the stripping of the British Crown of all it held so dear in the United States. However, much we may condemn, from the standpoint of our own sympathy, all and singular in their behalf, there is nevertheless more or less of mitigation of wrong and treachery that he was doing it all in behalf of the country and sovereignty to which he owed allegiance. The colonies were proposing to fight for mitigation of the wrongs of unequal and unjust taxation and the refusal to them of the right of representation in the British law making body and yet out of that little movement which I have shown was only for the mitigation of wrong, came this the great idea, so suddenly developed, of independence, and from it has come all the glory of a mighty and united country."

While William Hunter, an eminent Ohio writer, of the *Chilicothie Advertiser*, says:

"It give us pleasure to note that the battle of Point Pleasant is called the first battle of the Revolutionary War by those who are celebrating the anniversary, although questioned by New England historians who seem to believe that the whole war was fought in a radius of twenty miles of Boston, and the most has been made of every little skirmish in that region, while the battles in the Western country are not even mentioned."

State Historian V. A. Lewis again says, in the *West Virginia Historical Magazine*, of the battle.

"It is the greatest event in the colonial period and stands just at its close. With it the Revolutionary Period begins. Hence the battle is as it were the connecting link between two of the great periods in all American History. Closing as it does the one, and opening the other.

Edward Ingle, writing in the *Manufacturer's Record*, in November, 1901, on the Preservation of Virginia's Antiquities, says:

"Andrew Lewis, not a Virginian, but yet a type of the rearguard of the Revolution, fought successfully at Point Pleasant in 1774, that which was really the first battle of that struggle and a battle far reaching in its significant results."

While a bill introduced by Senator Scott passed the Senate in 1905, carrying an appropriation, Senator Scott wrote and offered his personal check for one thousand dollars with which to erect the monument if the commission would abandon the idea of securing aid from the National Congress. A hasty conference of friends of the movement was held and the offer declined, as the agitators of the monument building were not only anxious that the Government should appropriate adequate funds, but that it should officially recognize the battle as one of those of the Revolution; so the offer of Senator Scott was politely declined. Congressman Hughes pressed the passage of the bill at that time in the lower house of Congress. Hon. J. T. McCleary, Chairman of the Committee to which the bill was referred, wrote a letter to Mr. Hughes, as follows:

"As I advised you yesterday, the committee adopted a policy more than a year ago as its policy for this Congress, that of making no appropriations for monuments to be erected outside of Washington."

The monument commission assisted by Mrs. Poffenbarger, then hastened to Charleston, where the Legislature was then in session, and the *Charleston Mail* gives, in part, the following:

"Mrs. Livia Simpson-Poffenbarger arrived here Thursday from her home at Point Pleasant, to aid in securing an appropriation for a Point Pleasant Battle Monument. She has attended a part of two former sessions of the Legislature for the same purpose. When seen by a *Mail* reporter in the office of her husband Judge George Poffenbarger, of the Supreme Court, she said in response to the question, when asked what the prospect for an appropriation is?"

"It is the first time I have ever believed we would get an appropriation when it was asked for. I believe now we are going to get it. We expect to get a recommendation for an appropriation through the Joint Finance Committee, and if we get a favorable report from the committee we have "crossed the Alps'" for there is positively no opposition to it this year outside of whether or not there will be available funds."

"What amount have you asked for, Mrs. Poffenbarger?"

"I had a most courteous hearing before the committee who had previously heard the members of the Monument Committee and I asked for $25,000,00, payable $5,000.00 annually."

"But can the state make an appropriation Governing five years?"

"No, but they can for two and the appropriation may be for a monument not to exceed $25,000.00 expense to the state of West Virginia, $5,00.00 of which is available now and $5,000.00 next year, and the rest may be implied, as in the case of the appropriation made for the West Virginia Hospital at a cost of $80,000, $10,000.00 of which was available when appropriated."

Here followed the history of the effort made for the erection of a monument.

Mrs. Poffenbarger telegraghed her paper –

'Charleston, W. Va.,

February 28, 1905.

The State Gazette

The Senate at 6:30 P. M. passed an amendment to the Appropriation Bill of $2,500.00 for this year and $2,500.00 for 1906. The hill still has to go back to the Mouse for concurrence and if it tails there, will commitee."

The amendment was offered by Senator E. S. McCown, of the Fourth Senatorial District. His speech was one of his best efforts while Senator Darst worked heart and soul for the appropriation. The amendment failed to pass the house, but was saved in the Joint Conference committee where by herculean work the building of a battle monument at Point Pleasant was assured. The state had once again assumed the responsibility and every year since the Legislature has appropriated money with which to build the monument. Except this appropriation for 1906 and 1907, all has been vetoed except that of $1,000.00 made in 1909, because of lack of funds as announced by Governor Dawson.

Feb, 15, 1906, Hon. James A. Hughes, who was pressing Congress for an appropriation for funds with which to build the monument wrote as follows from Washington:

Mrs. Livia Simpson-Poffenbarger

Point Pleasant, W. Va.

I am in receipt of a letter from Mr. Austin who states that it is the opinion of the Monument Committee of your place that the amount asked for be reduced from $50,000.00 to $10,000.00."

"The amount carried in the bill does not amount to anything as the committee would only appropriate such an amount as they saw fit and would be governed by the wishes of the committee. I will introduce another bill carrying $10,000.00 instead of $50,000.00 as in the present bill. I had a talk with Mr. McCleary, and he told me that the committee still had under consideration the advisability of whether they would make any appropriations outside of the City of Washington. So far, they have not come to any conclusion. I had a talk with Senator Scott in regard to this appropriation and he doubted very much whether they would make any appropriation outside the City of Washington and he advises that this monument should be erected and that it should be done by private subscription and in addition to what the State had already appropriated he said he would be glad to head the list with a private subscription."

"Now I want to advise you frankly about this, if the Library Committee of the House refuses to make any appropriations for monuments, outside of the City of Washington, I think it will be useless to press the matter further, and I think it would be well to consider the suggestion of Senator Scott. I will be glad to hear from you in reference to the matter.

Very truly yours,

J. A. Hughes."

To the above letter, Mrs. Poffenbarger replied:

"We do not ask that the Congress of the United States build the Battle Monument at Point Pleasant because the funds cannot be raised by private subscription or secured as an appropriation from the State of West Virginia, but because we want the Government to officially recognize the batttle as it was in truth a battle of the Revolution, indeed, the First Battle of the Revolution, and no matter how insignificant the approation, if the bill correctly states its status we will be content to raise the money necessary as best we can, although we want as large an appropriation as we can get. While we appreciate the generosity of Senator Scott, should he donate the entire amount necessary it would fail in our main purpose of having the government officially credit the battle the honor it deserves and we will have again to decline his offer and insist that you both press the matter before Congress so vigorously as to ultimately bring the desired result. Again thanking you and Senator Scott for your past efforts and expecting renewed zeal, I am

Very truly

Livia Simpson-Poffenbarger."

That the Congress of the United States was still importune is evidenced by the fact that on December 4, 1907, Senator N. B, Scott introduced Senate Bill 160 which was favorably reported February 17. 1908, without amendment, as follows:

"A BILL to aid in the erection of a monument or memorial at Point Pleasant, West Virginia, to commemorate the Battle of the Revolution fought at that point between the Colonial troops and Indians October tenth, seventeen hundred and seventy- four."

An identical bill was introduced in the lower House of Congress by Hon. James A. Hughes.

A Telegram, as follows, brought the first intelligence to Point Pleasant that the bill had passed both branches of Congress:

Mrs. Livia Simpson-Poffenbarger

Congress appropriated $10,000.00 for a battle monument at Point Pleasant. Congratulations.

James A. Hughes.

Washington, D. C

The Monument Commission got busy and realizing that the introduction of new methods of monument building, lasting as the pyramids of Egypt, had been introduced, by which the monument could be built with the money available, a contract was let for the monument at a cost at the factory of $15,000.00, the shipment and erection of which would make a total cost of $16,000. The monument to be built of Balfour granite, the statue thereon to be of Westerly granite. The shaft is an obelisk with a base twenty-four feet square, the height to be eighty-two feet. The statue is to be that of a colonial soldier of the primitive Virginia style, dressed in hunting shirt, coon skin cap, leather breeches and long rifle. The whole to be completed for the unveiling of the monument on the One Hundred and Thirty- fifth anniversary of the battle of Point Pleasant, October, 10, 1909.

The Monument Commission on June 10, 1909, issued the following announcement:

"ATTENTION CITIZENS. A general invitation is extended for a citizen's meeting at the Court House at Point Pleasant on Thursday June 10th at 8 P.M. to make preparations for the celebration of the Battle of Point Pleasant, the unveiling of the monument and Home Coming Week, under the auspices of the Mayor and Civil Authority of the town of Point Pleasant, and interested citizens of the town, county and state.

Signed,

John P. Austin,

President of the Monument Committee.'"

There was a large and enthusiastic meeting held in response to the call. Judge John Lamar Whitten, Mayor of the town, who presided, was elected as the permanent chairman of arrangements for a West Virginia Home Coming Week, Celebration of the Battle and Unveiling of the Monument, October 7-8-9-10, 1909. The others appointed to further the success of the proper observance of the battle were the following Committees in charge of the Celebration.

ORGANIZATION

Chairman, Mayor John L. Whitten.

Secretary, E. Jacob Somerville.

Asst. Secretary, Warren C. Whaley.

Grand Marshall, Col. John P. R. B. Smith.

Assistants, Lewis C. Somerville, Peter Higgins Steenbergen, Edward Barto Jones.

COMMITTEE AT LARGE

Hon. John P. Austin, President Monument Commission; Monument Commissioners, Hon. V. A. Lewis and Mr. C. C. Bowyer; Mayor, Judge John L. Whitten. Col. John P. R. B. Smith and Mrs. Livia Nye Simpson-Poffenburger.

FINANCE COMMITTEE

Joseph Friedman, Chairman, Peter Higgins Steenbergen, Edward K. Thomas, Peter S. Lewis, Griff T. Smith, Tol Stribling, Hon. J. Samuel Spencer, John G. Stortz, Point Pleasant; R. J. Patterson, Maggie; William R. Thompson, Huntington; Charles Cameron Lewis, Jr., Ex-Governor Wm. A. McCorkle, Hon. Jno. O. Dickenson, Charleston; Ex-Governor A. B. Fleming, Fairmont.

INVITATION COMMITTEE

Governor Wm. E. Glasscock, Senator Stephen B. Elkins, Senator Nathan B. Scott, Hon. James A. Hughes, Judge Ira E. Robinson, Judge George Poffenbarger of West Virginia; Col. H. R. Howard, John E. Beller, Robert E. Mitchell, B. H. Blagg, Andrew Lewis Boggess, Charles C. Lewis, H. Green Nease, Hon. George W. Cossin, James M. H. Beale, Hon. J. Samuel Spencer Judge John W. English, Judge Wm. A. Parsons, Point Pleasant; Hon. E. O. Randall, Columbus, Ohio; Hon. W. Sidney Laidley, William Burdette Mathews, Charles C. Lewis, Sr., John Q. Dickenson of Charleston; Gen'l. Charles H. Grosvenor and Hon. Jerry Longfellow Carpenter, Athens, Ohio; Hon, Edmond Sehon, Dr. Lewis V. Guthrie and Louis Sehon Pomeroy, of Huntington.

TRADES DISPLAY

Robert J. Heslop, Charles K. Blackwood, Charles E. Jones, David S. Snyder, James Walter Winden, Mark Shiflet, Horton Roseberry, Joseph W. Rhoades, Will Filson, John C. Franklin, Ed. Lawhead, H. H. Henry, C. Frederick Hess, Captain C. Homer Varian, Robert Kiger, James Stephenson, George Miller, John Wells, Geo. W. M. Hooff, Alexander B. McCulloch, James Cavenaugh, James B. Tippett, F. B. Tippett. Lemuel Shiflet, Hugo Juhling, Jr., Bertram L. Burdette, Frank Fadley, Enos B. Thomas, H. W. Ellis, Dr. Ed McElfresh, Jacob P. Hetherington, L. J. Coley, Peter C. McDade, W. A. Williams, Harry M. Langley.

DECORATION AND SPEAKERS STAND

Frank Filson, B. Franklin, Jr., Charles Russell McCulloch, Wm. H. H. Gardner, J. Floyd Burdett, S. Waldo Swisher, Alonzo Walker, Enos, C. Winger, John W. C. Heslop, Ed A. Arrington, Edward W. Craig, Geo. P. Gardner, Wm. Tully, John Love, Sam'l Lutton, Fredrick Capes, Charles Dashner, G. E. Mathews, Mesdames Harry E. Burnside, Wm. Steenbergen, Robert E. Mitchell, Edward McElfresh, Rush H. Burnside, Edward J Burnside, Lem Shiflet, Samantha J. Baum, Wm. Steinbach, Mrs. M. Friedman, Mrs. Kate Williams, Mrs. J. W. English, Mrs. Robert P. Lynch, Mrs. Ella Fenton, Mrs. Horton Roseberry, Mrs. E. H. Woelffel, Mrs. Wm. Kenny, Mrs. E. H. Armstrong, Mrs. H. A. Barbee, Mrs. James B. Tippett, Mrs. Wm. C. Stortz, Mrs. J. F. Burdett, Mrs. B. Franklin, Jr., Mrs. George Comstock, Mrs. Joseph H. Holloway, Mrs. Ashabel Hughes, Misses Edith Tippett, Maud Kisar, Reba Beale, Hattie Price, Mary Lewis, Venie and Jessie Thomas, Mrs. Homer Smith, Mrs. E. B. Jones, Mrs. Wm. E. Hayman, Mrs. W. C. Whaley, Mrs. Tol Stribling.

ADVERTISING COMMITTEE

Homer Smith, Dr. W. P. Neale, Marcus Friedman, Warren C. Whaley, Charles C. Lewis, Edward C. Berridge, Dr. Ed. McElfresh, John F. Lewis and James B. Tippett.

ENTERTAINMENT COMITTEE

Griff T. Smith, Howard L. Robey, Lesley L. Neale, Geo. C. Somerville, Dr. Frank V. Butcher, Ed Filson, Wm. Steinbach, Howard Long, Geo. W. Long, John L. Hutchinson, E. H. Woelffel, W. W. Riley, R. P. Liter, Lem Shiflet Jr., Point Pleasant; John D. Lewis, Phil Walker, John Baker White, Charleston; Lycergus N. Knight Maggie; M. G. Tyler, H. E. Spilman, Dr.

Richard Stone, Spilman, Dr. Charles Petty, Hartford; Mesdames John Samuel Spencer, Charies Clendenin Bowyer, John L. Whitten, Charles C. Lewis, John Daniel McCulloch, H. Green Nease, John W. C. Heslop, L. J. Williamson, Rankin Wiley, John McCulloch, Frank Filson, Wm. P. Neale, Peter Higgins Steenbergen, Hiram R. Howard, J. H. Wade, Jackson Lee Pannel, Charles Russell McCulloch, James W. Windon, Patrick F. Ryan, John P. Austin, Robert P. Liter, Howard L. Robey, E. Barto Jones, Joseph Friedman, E. B. Sisler, B. Franlin, Jr., E. E. Thomas, Mary Margaret Bryan, Homer Smith, Ben Franklin, Sr., S. W. Swisher, Walter Lincoln; Misses Josephine Howard, Irene Bowyer, Lillie Lee Hogg, Elizabeth Harding Hogg, Julia Polsley, Ada Gilmore, Cornelia Smith, Veva Haptonstall, Margaret Lynn Neale, Gertrude Howard, Edith Tippett, Venie and Jessie Thomas, Kate Stribling, Lena L. Roseberry, of Point Pleasants; Mrs. Geo. W. Gist and Miss Maggie Hayman, Letart; Mrs. W. L. Lawson and Miss May Jackson, New Haven; Mrs. M. M. Brown, Mrs. D. E. Newton, Hartford; Mrs. Joseph H. Windon, Maggie; Mrs. V. A. Lewis, Mrs. W. E. Ruttencutter, Mrs. L. E. Bletner, Misses Maud and Annie Lewis, Mason; Mesdames John McCoach, Edmond Sehon, Columbus Sehon, Taylor Vinson, Wm. R. Thompson, Mary Lesage, Margaret Lynn Harvey, C. R. Thompson and James A. Hughes, Huntington; Mrs. Kate Sterrett, Mrs. Wm. H. Vaught, Mrs. John Thornberg, Five Mile; Mrs. M. Ella Hutchinson, Henderson; Miss Lizzie Smith, McCausland; Miss Francis M. Maupin, Arbuckle; Mrs. Charles E. McCulloch, Five Mile; Miss Rhoda Long, Mrs. Monroe Poffenbarger, South Side; Mrs. C. A. Green, Otia; Mrs. James Henderson, Five Mile; Dr. A. G. Martin, F. M. Middleton, Winfield; Dr. C. McGill, Red House; Robert Brown, O. F. Stribling, Apple Grove; John H. S. Spencer, Graham Station; F. C. Hute, John C. Levzey, L. Quickie, Thos. L. Finney, Pliny, Frank Dunn, South Side; George L. Sebrell, E. B. Nease, Arbuckle; Will Armstrong, Gallipolis; Dr. Blake, R. E. Blake, Henry McCoy, J. B. Frazier, Buffalo; Robt. Somerville, Maggie; W. W. Cornwall, Glenwood; Earl Henry, Clifton; A. G. W. Brinker, A. C. Cross, Thos. Z. Blessing, Letart.

SPEAKERS COMMITTEE

Col. H. R. Howard, Capt. W. H. Howand, Wm. E. Hayman, Judge George Poffenbarger, Robt. L. Hutchinson, Hon. J. S. Spencer, Benjamin Franklin, Sr., Enos B. Thomas, Dr. E. J. Mossman, Carlisle L. Whaley, Rev. J. H. Gibbons, Rev. Pullin, Rev. R. P. Bell, O. A. Roush, Geo. McClintock, Capt. Rush H. Burnside, Benjamin H. Blagg.

MUSIC COMMITTEE

E. B. Sisler, J. H. Norton. B. F. Gibbs, A. C. Van Gilder, Charles K. Blackwood, Lew McMillen, Rankin Wiley, John G. Aten, Wm. Steenbergen, Dr. John Fadley, Mesdames Mary Margaret Bryan, J. M. H. Beale, Carlisle L. Whaley, Eliza Waggoner, Charles Filson, Nannie E. Hale, Kossuth T. McKinstry, Misses Josephine Beale, Gertrude Howard, and Margaret Malone, Mrs. W. C. Whaley.

PROGRAM COMMITTEE

Lewis C. Somerville, Judge, George Poffenbarger, Col. Hiram R. Howard, Howard L. Robey, Robt. L. Hutchinson, Charles Buxton, Judge John W. English.

TRANSPORTATION COMMITTEE

Wm. C. Jordan, L. C. Kuhn, Moulton Houk, F. Gerald Musgrave, B. H Blagg, Wallace A. Barnett, John McCulloch Dr. Hugh A. Barbee, Wm. W. Bryan, Ed. C. Berridge, Capt. E. A. Burnside, Fred Smith, Capt. E, E. Varian, Capt. John Thornberg, Gus Fry, Jos. L. Ruth, Elmer Nutter, E. B. Martin, Capt. Gordan C. Greene, J. W. Hooper.

SCHOOL CHILDREN

Prof. H. E. Cooper of Point Pleasant and the teachers of Mason County.

ADVISORY COMMITTEE

Messrs. John W. Steenbergen, Beale; Wm. J. Keister, Ashton; Clinton Poffenbarger, Mason Long, Beech Hill; James W. Long, Hon. Jabez Beard, South Side; E. F. Bletner, W. E. Ruttencutter, Mason; Charles Juhling, Hon. J. M. Hensley, J. M. Chapman, D. E. Newton, Capt. M. M. Brown, Hartford; Judge W. W. Jackson, Geo. N. Capehart, W. L. Lawson, C. T. Bumgarner, New Haven; Geo. W. Gist, Dan Sayre, Wm. Klingensmith, Letart; Philip Click, Willow Tree, Judge Byrd Stone, W. P. Smith, Fred Sullivan, Wm. Jividen, Charles F. Thomas Leon; Judge W. H. Vaught, Robert P. Morris, Henry Fry, Henderson; James Henderson, Rankin Hill, Henderson; Hon. Jas. L. Knight, Messrs. Asa Musgrave, James W. Windon, Jos. H. Windon, H. J. Norton, Judge B. J. Redmond, Dr. A. R. Girard, Hon. Geo. Parsons, Pleasant Flats; Capt. L. S. Parsons, John R. Couch, Hearne. W. H. Sayre, Chas. W. Hogsett, Wm. H. Rowsey, C. A. Green, of Hannan District, Shepherd W. Moore, Elwell; Geo. W.

Pullin, C. G. P. Musgrave, Debby; Geo. J. Meadough, James W. Kindey, W. H. Clarke, J. W. Bryan, Ash Hughes, W. H. Zumbro, Point Pleasant; John McCausland, Jr., Jno, R. Couch, B. K Bell, Hearne, R. W. Bateman, S. A. McNiel, Mercers Bottom, Judge J. L. Thorne, Wyoma. Col. Jerome T. Bowyer, Winfield; B. J. Lerner, Hartford; John J. Dower, Letart; Jas. T. Edwards, Clifton; H. C. Turner, Mason; David Caldwell, Gallipolis Ferry; Judge A. M. Pugh, Col. John L. Vance, Columbus, Ohio; Gen'l. John McCausland, McCausland; James M. Nye, Marrietta, Ohio; Mrs. Samaria, H. Palmer; Athens Ohio; Miss Margaret Lynn Price, Lewisburg, Mrs. Miram Donnally, Mrs. E. W. Wilson, Charleston; Mrs. Sanders Johnston, Dr. Adeline E. Portman, Washington, D. C.; Mrs. Anna S. Greene, Culpeper, Virginia; Miss Mary C. Nye, Marrietta, Ohio; Mrs. Sophia Dale, Belpre, Ohio; Mrs. M. C. Scott, Pomeroy, Ohio.

The *State Gazette* of August 15th, gives the following:

"On Monday August 2nd, 1909, at 11 a. m. the steam whistle on Captain Charles Homer Varian's pumpboat, lying in the mouth of Kanawha river, sounded a glad cry that was lustily joined in by the many steamboats lying in harbor. Our people came out in great crowds to learn the cause, and the on coming tide of people were directed to Tu-Endie-Wei Park, where had just been set the apex stone that completed the stone work of the splendid Balfour granite monument, commemorating The First Battle of the Revolution, fought at Point Pleasant, October 10th, 1774.

It was an occasion that for many years has been devoutly wished for, and there were many upon the grounds prior to the blowing of the whistles, who for years had watched every step of the preparation for the monument building. Among them were Mrs. J. D. McCulloch, who was a member of the Ladies Monument Association, that put by the first contribution, which with its accumulations, represented $2,000 invested in the Monument; Mr. C. C. Bowyer of the Monument Commission, who have so faithfully labored in the cause entrusted to them by the State; and Mrs. Poffenbarger whose interest has never lagged, and it was her little son, Perry Simpson-Poffenbarger, who suggested, and induced Capt. Varian to start the whistles.

The monument has been erected so speedily that our people can scarcely believe it is so nearly completed. This is accounted for by the fact that the stones were cut and numbered and ready for placing, and needed little cutting when they came upon the grounds. The Van-Amringe Granite Company of Boston, are the contractors, with Mr. J. E. Amedon of Merchants Depot, of Vermont, as the superintendent of construction here. Captain Charles Fredrick Hess was the contractor, not only for the splendid cement walks and walls, but for the cement work of the Monument, and the great underfooting was laid prior to June 7th, when the first carload of granite reached here.

On June 9th, the corner stone was laid. There were no ceremonies attending it and no deposits made save that of a small coin of the issue 1909, the year of the Monument construction.

However, as is the custom in the erection of such structures, a box was deposited in this monument.

It was found that in the center tube in the top section immediately under the great cap stone that binds the building, there was room to admit a box three inches in diameter and twelve inches long. Filson Brothers were called upon to construct a copper box of these dimensions and make it air tight. In it were deposited an Industrial Edition of the *State Gazette* of the issue, of February 2nd, 1905, upon which was written the following inscription:

"Deposited, Monday, August 2d, 1909, the date of the setting of the cap stone of the Point Pleasant Battle Monument, by Nathan Simpson Poffenbarger and Perry Simpson Poffenbarger, sons" and Natalie Simpson Bryan, niece of Geo. and Livia Nye Simpson-Poffenbarger.

A copy of the diary written by Margaret Lynn Lewis, wife of John Lewis the emigrant and a founder of the city of Staunton, Virginia, was placed in the tube upon which was the following inscription:

"Deposited, August 2nd, 1909, the date of the setting of the capstone of the Point Pleasant Battle Monument, by Sallie Lewis McCulloch, (Mrs. J. D. McCulloch) Great, Great Granddaughter of Margaret Lynn Lewis and Great Granddaughter of Col. Charles Lewis. Sallie Lewis McCulloch (Mrs. P. H. Steenbergen), Great, Great Granddaughter of Col. Chas. Lewis."

An Indian arrow bead taken from the ground when the excavation was made, was put in the box and with it a slip of paper bearing the following":

"This Indian arrow head is deposited by C. F. Hess, contractor for the cement work of this monument. It was found when the excavation was made."

The most important deposit made however, was a copy of "The Battle of Point Pleasant," bearing the following inscription:

"Deposited, Monday, August 2d, 1909, the date of the setting of the cap stone of the Point Pleasant Battle Monument.

"The illustrations and last pages are omitted on account of the inability of the printer to finish the volume by the date of the completion of the monument. Livia Nye Simpson-Poffenbarger. The Author."

This was the most valuable because, though not quite completed, it carried the most complete roster of the participants of the battle ever as yet published, that of 1080 men who participated at Point Pleasant and were entitled to share in the honors of this victory.

When the last stone bad been set in place, Mr. Amedon presented Mrs. Poffenbarger the two remaining blocks of granite from which will be made souvenirs of the monument.

The statue which is to be placed on a base in the front of the monument has not reached here, nor have the eight bronze plates in bas relief, two of which bearing the coat of arms of the United States and of West Virginia, and six of which bear the inscription of the killed and wounded and the officers commanding the army, but they have been shipped and will be here to be put in place by Mr. Amedon upon his return from St. Louis, less than three weeks hence. Upon his return, the monument will be pointed up, the statue and plates set and the monument veiled ready for the ceremonial attendant upon the unveiling of October 9th, 1909.